Am I My Parents' Keeper?

Am I My Parents' Keeper?

AN ESSAY ON JUSTICE BETWEEN THE YOUNG AND THE OLD

Norman Daniels
Tufts University

New York * Oxford
OXFORD UNIVERSITY PRESS

Oxford University Press

Oxford   New York   Toronto
Delhi   Bombay   Calcutta   Madras   Karachi
Petaling Jaya   Singapore   Hong Kong   Tokyo
Nairobi   Dar es Salaam   Cape Town
Melbourne   Auckland

and associated companies in
Berlin   Ibadan

Copyright © 1988 by Norman Daniels

First published in 1988 by Oxford University Press, Inc.,
200 Madison Avenue, New York, New York 10016

First issued as an Oxford University Press paperback, 1989

Oxford is a registered trademark of Oxford University Press

Library of Congress Cataloging-in-Publication Data
Daniels, Norman, 1942–
Am I my parents' keeper?
Bibliography: p.
Includes index.
1. Distributive justice.   2. Intergenerational relations.
I. Title.
JC578.D36   1988   362.6′042′01   87-12235
ISBN 0-19-505233-1
ISBN 0-19-506164-0 (PBK.)

2 4 6 8 10 9 7 5 3 1

Printed in the United States of America

To my elders,
who have taught me about the lifespan:
For Evelyn and Bess,
In memory of Monte, Fanny, Bertha, and Nathan

Preface

This is an essay about the just distribution of resources between the young and the old. It seeks a principled way, rooted in a theory of justice, to resolve disputes about how income support, health care, and other social resources should be allocated to different age groups in our society. This book is primarily concerned with the just design of social institutions, but it has a bearing on the dilemmas faced by individuals trying to understand what they ought to do for their elderly parents. Although it addresses important issues of public policy and, therefore, is of interest to all those concerned with the well-being of the elderly, current and future, it is a philosophical essay rather than a detailed analysis of policy options.

The central ideas for this book first came to me while I was working on my earlier book, *Just Health Care* (Cambridge University Press, 1985). Health care was of special moral importance, I argued, because it protected equality of opportunity for individuals. An obvious objection to this view was that the "opportunity" of the elderly lay in their past, which implied that there was little reason to provide them with health care. I had to reply to this objection, and thus I was led to think about what age bias in general involved. I came to understand that the age-group problem was a distinct problem of distributive justice, and I tried to articulate its relationship to other work in the theory of justice.

A personal experience involving my aged great-aunt provided me with an important insight into the problem. When my great-aunt became quite frail and suffered some mental impairment, her daughter found it extremely difficult to get adequate home care or an appropriate nursing-home placement. When acute episodes threatened her life, however, she was rushed to intensive care facilities and exhaustive efforts were made to extend her life. In the final of these episodes, I was reassured by her daughter that "the doctors are doing everything to save her." I suggested that perhaps it was time to let her die peacefully, but I was rebuked. "It's my mother—I can't do that." I then asked my cousin whether she would want her daughter to treat her in the same way she was treating her mother. "God forbid," she said, "when my time comes, I just want to go."

I was struck by the contrast between my cousin's view of how she would want to be treated and how she felt compelled to treat her mother. Her own view of how she would want health-care services allocated to her within her own life gave her no guidance in her decisions concerning her mother. Of course, how we would want to be treated is not always a good guide to how we ought to treat others, the Golden Rule notwithstanding. Still, it seemed to me that my cousin's considered preferences about how she wanted to be treated should have more to do with how her mother was treated than the design of our health-care institutions allowed. I began to explore the idea that the socially prudent design of our health-care and income-support institutions should be our guide to what justice requires in the treatment of the elderly. That idea became the central theme of this book.

Though an essay in philosophy, this book is aimed at a diverse audience of nonphilosophers as well. It is relevant to the thinking of sociologists, economists, and gerontologists who have been studying the varied aspects of this "Age of Aging," as Abraham Monk has called it. It provides a unified perspective through which public policy advocates, planners, and administrators can think about proposals for social reform, and it discusses issues of considerable concern to health-care providers concerned with the problems of long-term care or facing attempts to control health-care costs of the elderly. It will be understandable to undergraduates in courses on moral issues, public policy, or aging.

I believe the book will also be of use to the general reader—to anyone who is troubled by the widespread talk of the old and the young competing for scarce resources in our aging society and who wants to know what ought to be done. When public policies concerning the young and the old are improperly or unjustly designed, it not only hurts each of us during the various stages of our lives, but it also strains our ability to meet our family obligations. I provide motivation for my approach to these problems in Chapters 1 and 2, and I develop its theoretical core in Chapter 3. In Chapters 4, 5, and 6, I apply my Prudential Lifespan Account to health care, first in general, and then to the problems of rationing acute care and providing for long-term care. In Chapter 7, I apply the account to income support and Social Security, and in Chapter 8, I discuss some reform or "half-way" measures. In the Appendix, which will be of particular interest to philosophically inclined readers, I discuss recent criticisms of some central assumptions about the nature of prudence and the moral importance of respecting persons which underlie my approach.

My work on this subject was first encouraged by Daniel Wikler, then (in 1981) Staff Philosopher for the President's Commission for the Study of Ethical Problems in Medicine. He commissioned the paper titled "Am I My Parents' Keeper?," which appeared in several versions and ultimately as Chapter 5 of *Just Health Care*. Since 1983, through generous support from the Retirement Research Foundation, the National Endowment for the Humanities, and Tufts University, I have had extensive time released from teaching to research the material contained here. I could not have written this book, or the papers on which it is based, without this institutional support.

I owe important debts to many individuals as well. I especially want to thank Dan Brock, Joshua Cohen, John Rawls, and Dan Wikler for the many hours they have spent discussing these issues with me when I most needed a critical audience. Dan Brock and Allan Buchanan have generously provided me with detailed written comments on the whole manuscript. The following people have also helped me through discussion or written criticism of my work: Jerry Avorn, Hugo Bedau, Christine Bishop, Margaret Carter, Judith DeCew, Daniel Dennett, Leslie Frances, R. M. Hare, Andrew Reschovsky, Christina Hoff Sommers, Lawrence Stern, and Stephen White. My wife Anne has read various parts of the manuscript and has offered steady encouragement. I especially want to thank Constance Putnam who has provided helpful editorial comments on the whole manuscript.

Some parts of this work draw on previously published papers as follows:

"Family Responsibility Initiatives and Justice Between Age

Groups." *Law, Medicine, and Health Care* 13(4)(September 1985):153–159.

"Why Saying No to Patients in the United States is So Hard: Cost-Containment, Justice, and Provider Autonomy." *New England Journal of Medicine* 314(May 22, 1986):1381–1383.

"Equal Opportunity, Justice, and Health Care for the Elderly: A Prudential Account." In Stuart F. Spicker, Stanley R. Ingman, and Ian R. Lawson (eds.), *Ethical Dimensions of Geriatric Care*, 197–221, Dordrecht: Reidel Publishing Co., 1987.

Medford, Mass.   N.D.
April 1987

Contents

Am I My Parents' Keeper?

1

Conflict Between the Old and the Young

Problems of an Aging Society

We bathe in the Fountain of Old Age, even as we continue to search for the Fountain of Youth. Never before has there been such a high percentage of people over age 65, 75, or 85. Indeed, those over 75 and 85 are the fastest growing age groups in the country. Never before have so many adults planned for an old age they will actually live to experience. This aging of society is not just an artifact of the baby boom and baby bust of recent years, though these have their impact. Major, long-term social investments in public health, improved nutrition, and—with lesser effect—medical services have lowered death rates. More important, complex socioeconomic forces have drastically reduced birthrates in industrialized societies through most of this century. In fact, the United States is behind some European countries, where the aging of society is more pronounced. The Fountain of Old Age is fed by deep springs not likely to run dry soon.

The aging of society forces major changes in the institutions responsible for social well-being. As the "age profile" of a society—the proportion of the population in each age group—changes, social needs change.[1] As society ages, proportionally fewer children need education,

1. See Russell (1982) for a detailed study of the effects of the baby boom generation as

fewer young adults need job training, but more elderly need health care and income support. Changing needs find political expression. Strong voices press for reforms of the institutions which meet these needs. At the same time, advocates for existing institutions and their beneficiaries resist change. The result is a heightened sense that the old and the young are in conflict.

Specifically, there is a growing perception that the old and the young are locked in fierce competition for a critical but scarce resource, public funds for human services. The scarcity of these funds, largely the result of recent political trends, only sharpens the competition. In this context many who complain that the old are benefiting at the direct expense of children, the poor, or younger workers are crying "foul." They rally around a call for "generational equity." Indeed, there is now a Washington lobby called Americans for Generational Equity (AGE).

The Old Versus Children and Other Groups

Perhaps the most plaintive cries are about competition between the elderly and children, between grandparents and great-grandparents and their grandchildren. Since 1970, for example, a significant expansion of Social Security benefits has reduced poverty among the elderly—from double the national incidence to a level slightly below the average rate of poverty. In contrast, programs aimed at poor children, such as Aid to Families with Dependent Children (AFDC) have been cut back dramatically, especially since 1980, and many fewer poor children are elegible for help.[2] There are now proportionally more poor children than poor elderly, whereas fifteen years ago these ratios were reversed.[3]

Children who lose entitlements to AFDC lose more than income support for their families, since eligibility for other programs is affected as well. For example, children are losing in head-on competition with the elderly for Medicaid funds. They are now receiving a smaller pro-

---

this demographic bulge has passed through schools, entered the work force, and faces retirement.

2. Preston (1984:437) cites a Children's Defense Fund (1984a) calculation that in 1979 there were 72 children in AFDC programs for every 100 children in poverty, but there were only 52 per 100 in 1982. Preston cites Danziger and Gottschalk (1983:746) for their calculation that 56 percent of the elderly would have been in poverty in 1978 had it not been for Social Security income transfers.

3. The "incidence of poverty among children under 14 in 1982 is 56% greater than among the elderly, whereas in 1970 it was 37% less" (Preston 1984:436).

portion of Medicaid dollars than in earlier years, despite an increase in the number of children eligible.[4] This competition may be showing up in mortality statistics. Cutbacks in federal and state support for prenatal maternal care programs have been linked to some increases in infant mortality rates (Knox 1984), whereas increased life expectancy for the elderly coincides with heavy investments in Medicare.

If we look at aggregate measures of the competition between the old and children for all federal funds, we get a similar picture. In 1971, we spent less on the elderly than on national defense, but in 1983 we spent more, totaling $217 billion or $7,700 per elderly person.[5] In contrast, expenditures for all child-oriented programs—AFDC, Head Start, food stamps, child nutrition, child health, and all federal aid to education— totaled about $36 billion in 1984, about one-sixth of federal expenditures on the elderly. Federal per capita expenditures on children are only 9 percent of per capita expenditures on the elderly.[6]

In this competition there are not necessarily any bad guys. The elderly and their advocates are not villians, and their motives are no less pure than the motives of those who speak up for children. Still, in this sharp competition between grandparents and grandchildren, the children, at least the poor ones, are definitely losing.

The competition between the old and children for federal funds is sometimes seen as part of a larger competition, between the old and the poor. One advocate of "generational equity" complained that we spend four and a half times as much on the federal government's retirement programs, including Medicare, which benefit rich and poor alike, than we do on all welfare programs aimed specifically at the poor. We cannot help the poor, so the argument goes, if, as in 1985, about 28 percent of all federal spending goes to 11 percent of the population that is 65 and older—regardless of whether it is rich or poor.[7]

4. Preston (1984:437) cites a Children's Defense Fund (1984a) calculation that "Children's share of Medicaid payments dropped from 14.9 percent in 1979 to 11.9 percent in 1982 despite a rise in the child proportion among the eligible."

5. Preston (1984:438) cites the U.S. Bureau of the Census (1983a:343) for these figures.

6. Preston (1984:438) cites the Children's Defense Fund (1984b, Appendices) for these figures. It should be emphasized that federal figures for spending exaggerate the "bias" (if that is what it is) toward the elderly because they do not include the bulk of educational expenditures, which are financed at the state and local level.

7. See Longman (1985:73) for a clear statement of the complaint that this spending pattern involves generational inequity.

Another group widely portrayed as losing out to the elderly is the working-age population. Those who rally around the call for "generational equity" complain that the elderly are reaping "windfall profits" from Social Security and Medicare programs. For example, they typically collect benefits costing over three times as much as the taxes they contributed to these schemes. Of course, the working-age population benefits from some of these transfers to the elderly—they might otherwise have to contribute more private support to elderly parents. And the working-age population may think that they will receive generous benefits in the future, when retirement comes. But as Longman (1985:73), who is associated with the Washington lobby, Americans for Generational Equity (AGE), urges, "There may be a point at which the young say 'enough' and rise up in revolt against their elders." His point is that the 75 million members of the baby boom generation will not be able to spread the burden of their retirement over as large a group of workers as the current elderly now do. Each retiree is now supported by 3.4 workers, but when the baby boom retires, only two workers will contribute for each retiree.

The threat is to the long-term solvency of the Social Security and Medicare systems. One way to produce solvency would be to stimulate robust economic growth, but such growth would require extensive investment to promote capital formation and to enhance the productivity of the young. Such investment is impossible, AGE complains, when we invest so heavily in consumption by the elderly (Longman 1985:81). Other proponents of "generational equity" advocate a different strategy. They urge "privatizing" our system of income support for retirement by shifting workers away from the Social Security system and into individual or private retirement schemes which are based on vested assets, not transfers from the young to the old. In this way, each cohort will depend on its own resources and be responsible only for its own well-being.

Elderly Parents Versus Adult Children

The conflict between the elderly and other age groups over public money sometimes penetrates right to the level of the family. Currently, families provide about 80 percent of all home health care to the partially disabled elderly. A large number of people need such help. Some 3 percent of the noninstitutionalized elderly are bedfast and an additional 7 percent are homebound (Shanas 1979, cited in Frankfather et al. 1981:6). This care is costly to adult children and other family members, in terms of money, expended time, and the stress of sustaining care over extended periods. As one daughter put it,

> We put her in the shower and brush her teeth . . . . We put lipstick on her and put her hair in a French twist with floral combs . . . . When we go away overnight, we take her with us . . . . I put paper on the floor 'cause she doesn't always make it to the bathroom . . . . I'll paint her apartment, change the light bulbs, and wash the windows. (Frankfather et al. 1981:1)

Despite the high level of care provided by families, rising public expenditures on long-term care have led some legislators in the early 1980s to propose "family responsibility initiatives" in about half the states. If passed, such laws would hold family members legally responsible for costs currently paid by Medicaid. The effort is to shift costs out of public budgets, where financial responsibility is spread over the tax base for the program, and onto individual families wherever possible. Even if the overall costs of care for the elderly are not reduced through such initiatives, since individual families will still pick up the tab, the burden of bearing those costs will fall much more heavily on a narrower group of the young, those with frail elderly parents. For them, the burden of caring for elderly parents will intensify. Adult children will find that resources they need for their own impending retirement, or that they might spend on their own children's or grandchildren's education, will have to be spent on care for their own aged parents. Shifting costs out of public budgets, however, will not eliminate competition between the elderly and the young for resources. It will only shift the locus and burden of that competition—from public budgets to family budgets. Thus a public decision not to treat the provision of long-term care as a social obligation would only intensify conflict within families.

Unmet Needs of the Elderly

There is, of course, another side to the competition between the old and the young. Though we spend heavily to meet the needs of the elderly— and we have just seen arguments suggesting we spend too heavily—

their needs are far from met by existing institutions. Perhaps the clearest example of the failure to meet important needs is our long-term care system—the nursing homes, personal and home care services, and mechanisms for financing them. Of course, the long-term care system treats the young as well as the old, and by no means do all the elderly use it. It is important to remember that the elderly are heterogeneous, with regard to both family situation and health status. Relatively few elderly in the 65–75 age group suffer disability, but the frequency of disability increases dramatically with advanced age.[8]

The long-term care system is widely criticized for leaving millions of partially disabled elderly unable to sustain their normal lifestyle. People with only mild disabilities, who could maintain normal patterns of living if they had modest help, cannot find or finance the home care and social support services they need. For example, there is no market in which we can buy adequate private insurance for such care, and these services are usually not reimbursable with public money. Ultimately, millions of elderly are forced into premature and inappropriate levels of dependency on their families or institutions. Obviously, this failure poses problems not just for the disabled elderly, but for their families. Families who provide care must do so without the benefit of services, such as day-care or drop-off centers for the elderly, which provide temporary relief from the burdens of care. This is the part of the health-care system which we have left most in individual hands, without benefit of group risk-sharing or social insurance schemes.

Despite these inadequacies in meeting the needs of the elderly, the long-term care system is very expensive. State budgets, which contribute to Medicaid costs for nursing home care, groan with the burden. What is worse, since the most rapidly growing age groups are those over 75 and 85, the groups which need the most long-term care, there is no possibility of meeting future needs at today's levels without proportionally greater budgets for long-term care. It is thus likely that unmet needs will grow significantly.[9]

8. Soldo and Manton (1985), in their excellent study of health needs of the oldest old, report that 6.7 percent of those 65–74 require personal care assistance, 15 percent of those 75–84, and 44 percent of those 85 or over.

9. Some recent changes in the acute care system may exacerbate these problems. Medicare cost-containment measures reimburse hospitals a fixed amount that depends on the diagnostic category for each elderly patient. This gives hospitals strong incentives to release patients earlier than has been customary, but there is a lack of home care for such patients.

Not all health-care needs have been created equal. The vast reservoir of real, unmet needs for long-term care contrasts sharply with the extensive provision of acute-care services. Despite recent cost-containment measures affecting Medicare, we lavish life-extending resources on the dying elderly as if we were meeting their most urgent medical needs, and we point to this glamorous challenge to death as if it proved we value highly every last minute of their lives. And indeed, it is the "last minute" we appear to value most, as that is the one for which we most vigorously provide. But it is far from obvious that prolonging the process of dying in these ways meets an important health-care need. Often, such care merely traps the elderly in treatments they and their families do not want. Even explicit preferences the elderly may have to discontinue treatment are frequently ignored. Perhaps even more common is a trap we create out of obligations we feel to do everything possible to "save" our parents or our spouse, even if we would reject such treatments for ourselves had we the choice.

In sum, our health-care system reflects serious confusion about which of the health-care needs of the elderly are most important to meet. It behaves as if the most urgent need of the elderly is to forestall death when it is imminent, whatever the cost. But in the course of doing so we leave unmet urgent needs that plague millions of elderly for long periods of their lives, while they age. Were the system not lubricated by good intentions, it would seem open to the charge that it is biased against the elderly. That is, it appears to be designed to meet the needs of the young, whose profile of needs is quite different. Of course, the system is not lubricated only by good intentions, since it may also serve the interests of the providers of care better than the interests of elderly recipients, and these interests of providers may happen to coincide with the interests of the young.[10]

We have already seen that the proportion of the elderly population in poverty is comparable to that of the population as a whole, not any higher. But this still means that millions of elderly live in poverty, and millions more live near the poverty line. Moreover, the poverty of the elderly is unevenly distributed, for example, by race and sex. The needs of the elderly for income support thus remain substantial. We can reduce expenditures to meet these needs only by vastly increasing the number of elderly people living in poverty—unless we can target our transfers to the poor elderly more effectively, for example, through

10. I am indebted to Allan Buchanan for this point.

programs whose eligibility requirements are based on economic and not age criteria. Yet it is arguable that more poor elderly have been helped by the relative invulnerability of age-based programs than would have been helped had programs been more narrowly targeted for the poor. As we have seen, it is the programs aimed at the poor that have been cut back the most during periods in which public funds for human services are under attack.

Ironies of Success

A deep irony lurks in the conflict and competition we perceive between the young and the old. The conflict emerges as the result of success in pursuit of a long-standing social goal. We have long thought that the elimination of early death, the introduction of family planning to reduce pressures on resources, and the provision of social programs that meet basic needs throughout the lifespan would lead to greater well-being and a more harmonious, long life for all. Instead, the aging of society is seen as a demographic shift that pushes us more quickly toward resource scarcity. This has evoked a Malthusian forecast of doom: "The elderly produce little, yet their extensive needs give them large appetites for social resources. Their growing numbers make politicians eager to satisfy that appetite. But shifting resources from the young to meet those needs will further undermine productivity, and so competition will occur under conditions of even greater scarcity." This Malthusian panic contains its own, novel irony. In our aging society, the resource scarcity it claims we face is intensified by *dropping* birth- and deathrates in industrialized societies, not by the high birthrates which traditional Malthusians claim will push us beyond the resource limits of the planet. (Every old plot needs a new twist or it ceases to be entertaining.)

That problems in our aging society are the result of the success, not the failure, of longstanding social policy does not make them easier to tackle. They are especially difficult problems of public policy because it is the *basic needs* of different groups that lead to conflict and competition. The old and the young both need health care, for example. This is not a case in which the basic needs of blacks are not met while whites enjoy special privileges or affluence. In such a case we can justify redistribution by appealing to the greater urgency of the needs of the blacks. But what do we do when the competition between the old and the young is for life-extending health care or for income support necessary to meet basic needs? What do we do if the competition means a choice between health care for the elderly and education for their grandchildren? What do we do when the competition means a choice between the immediate needs of elderly parents and the future needs of their adult children? We need not be in the grip of Malthus to sense the poignancy as well as the difficulty of these questions.

Underlying the common perception of competition between the old and the young, underlying the call for "generational equity" in our aging society, there lurks a challenging "new" problem, which is the subject of this book. What *is* a just or fair distribution of social resources among the different age groups competing for them? I shall argue that we can solve this problem only if we think about the competition between age groups in a radically different way. Indeed, we will have to stop thinking about competition between groups altogether. But I get ahead of my story.

A "New" Problem of Justice

I have suggested that we face a "new" question of justice, "How should we distribute social resources among the different age groups competing for them?" It is this question that must be answered if we are to solve the problems posed by the aging of society. Yet this question about justice between age groups is not really a new problem, even if it suddenly seems urgent and even if it has been little discussed. Moreover, it does not arise merely because of the recent aging of society. Rather, the just distribution of resources between the old and the young is a problem for every society, past and present.

Each society has some distribution of resources among its distinct age groups, whatever its age profile. Each society has institutions and social practices that transfer income, power, and other social goods from one age group to another. Instead of merely describing such patterns, as social scientists do, we might search for the normative principle which, at least implicitly, governs such a distribution. Such a principle can be thought of as a particular society's answer to the question, "What is just between age groups?" It characterizes that society's conception of justice between the young and the old. We might then debate the merits of alternative conceptions of justice—alternative distributive principles—seeking the best answer to the question about justice between age groups.

Philosophers have paid little attention to this question, despite its importance. The issue has been ignored because moral questions are often camouflaged by the status quo. That is, we notice something is a problem only when traditional solutions to it break down, when long-standing institutions and patterns of behavior are disrupted. The aging of society has disrupted these patterns, and that is why the question I pose about justice between age groups appears new.

There is another reason why the question about age groups appears new. It is primarily in this century that society has assumed direct responsibility for transfering certain goods and services from the young to the old. When the responsibility for these transfers fell primarily within families, as it still does for long-term care, the age-group problem seemed to be an individual or family one and not a matter of social justice at all. It was couched in terms of family obligations, not requirements of justice, though the appeal to family obligations can be thought of as a particular answer to the underlying question about justice between age groups. Still, until the aging of society sharpened conflicts about the social obligations to transfer these resources, the age-group problem remained hidden.

Distinguishing Age Groups
from Birth Cohorts

In what follows the term "age group" will refer to people who fall within a certain age range or are at a certain stage of life. For the most part, I will be concerned with justice between the young and the old. I will often simplify the problem by designating those over 65 to be the *old* (or the *elderly*) and designating those of working age, between 16 and 65, to be the *young*. Referring to the old in this way involves a conventional if arbitrary boundary; referring to the young in this way stretches our ordinary usage a bit. Sometimes I will want to distinguish children and the "old-old" as age groups at the extremes.[11] Even if I mainly talk about justice between the old and the young, I seek a

11. The term "old-old" is not intended to pick out a group merely by age; the contrast with the term "young-old" is intended to be largely functional. See Neugarten (1974).

solution to the age-group problem that extends to other age-group distinctions.

The elderly are an age group including people over age 65. But the current elderly also belong to a particular birth cohort, namely, the cohort of people born prior to 1920. We may also say that the elderly are "the older generation." But "age group," "birth cohort," and "generation" mean different things. To answer the question about justice between age groups, we must distinguish these notions.

At a given moment, people in a particular age group also belong to a particular birth cohort. Over time, an age group comprises a succession of birth cohorts. Twenty years ago, the elderly included only pre-1900 birth cohorts. Today, they include all pre-1920 birth cohorts. Age groups do *not* age. Over time, new and different birth cohorts simply move into an age group. In contrast, birth cohorts *do* age. They pass through the stages of life, and so, at different times, fall into different age groups. The 1900–1920 birth cohorts were among the young in 1965; now they are among the old.

A birth cohort is a distinct group of people with a distinctive history and composition. The question, "What is a just distribution of social goods between birth cohorts?" thus carries with it the assumption that we are focused on the differences between distinct groups of people. For example, special questions of fairness may arise because of particular facts about the socioeconomic history and composition of particular birth cohorts. The notion of an age group abstracts from the distinctiveness of birth cohorts and considers people solely by reference to their place in the lifespan. Consequently, our question about justice between age groups also abstracts from the particular differences between the current elderly and the current young that arise because of the distinctive features of the birth cohorts which happen to make up those age groups. We are concerned with a common problem about justice between the old and the young that persists through the flux of aging birth cohorts.[12]

12. This is *not* to say that in general, for example, in the social sciences, we can talk about the old and the young in abstraction from the different experiences these groups have which derive from differences between cohorts. The experience of old age, for example, will vary from birth cohort to birth cohort, depending on facts about each cohort's education and prior history. See Featherman (1983).

Age Groups Versus Cohorts

It is tempting, in our actual roles as voters or public planners or tax-
payers or adult children or elderly parents, to think only about justice
between the current young and the current old. After all, each of us
identifies with one group or the other. It is easy, that is, to lose sight of
the more abstract, timeless question about age groups and to slip into
the more immediate question about justice between particular birth cohorts. How is *my* cohort faring under these policies? Will my cohort—
will I—do as well with Social Security or Medicare as the current
elderly? How will "we" do compared to "them?"

The question about justice between birth cohorts is an important
one—indeed it is foremost in the minds of many who ask for "generational equity," as we have seen. I will argue that it is not as central as
the age-group problem. This claim about the priority of the age-group
problem is controversial and requires a defense. Indeed, some will insist
that there is never any question of justice between the young and the old
except one that is about justice between particular cohorts—*these*
young and *these* old. They eliminate the age-group problem in favor of
the birth-cohort problem.

I think there are two moral problems, not one. Insisting, for example,
that different cohorts should be treated equitably or fairly does not tell
us just what transfers society ought to guarantee between the young and
the old. Knowing that what we do for one cohort must be equitable
compared to what is done for another does not help us to learn what we
should do for either as they age. Answering the age-group question
properly, however, may teach us what to do for each birth cohort over
time.

Another point suggests that these problems are distinct. The question
about age groups is centrally connected to certain other issues of justice
in a way that the question about birth cohorts is not. For example,
worries about age-bias and age-discrimination abstract from any consideration about birth cohorts. In asking whether people over 65 should be
required to retire, or whether they should be denied access to life-
extending medical services such as dialysis, as they are in Great Britain
(Aaron and Schwartz 1984), we are not asking a question which in any
way turns on differences between birth cohorts. We are asking a question about the treatment of different age groups. Similarly, other moral
issues, such as questions about filial obligations, are raised about the

young and the old in general. These issues too abstract from questions about particular birth cohorts. These points of difference between the problems motivate my decision to give priority to the age-group problem and to treat the special issue of justice between birth cohorts as a problem to be addressed after we understand what justice between the young and old involves. (In Chapter 7 I shall discuss the priority of the age-group problem.)

The term "generation" is ambiguous in several ways. A question about "justice between generations" can thus mean any of several things. A generation may be equivalent to a birth cohort, as in "the generation born in the 60s." Much of the discussion of "generational equity," which we encountered when surveying perceptions of competition between the young and the old, was really a discussion of equity between birth cohorts. We may also mean by "generation" groups of cohorts which do not exist at the same time. This reference to nonproximate cohorts is intended when we discuss the problem of obligations to future generations, which is sometimes called the problem of justice between generations. Thus we might ask what current generations owe to future generations with regard to preservation of the environment. Finally, we may mean by "generation" what I have called an age group. Thus, if we are concerned with the perennial stuggle between generations, we are talking about the conflict that persists over time between the old and the young, through the succession of birth cohorts. I shall restrict my use of "generation" to contexts in which we talk about justice to nonproximate, future (or past) generations, and I shall pay little attention to this problem of justice. It is worth noting that the kind of competition between age groups which concerned us earlier does not arise in the problem of justice between generations. Age groups coexist, cooperate, and compete in the same political and moral setting; future generations are at the mercy of current ones.

The question about age groups poses a distinct problem of distributive justice, though it is not, by any means, the whole of the problem of distributive justice. Any answer to it, including the one I develop, will consequently be only a fragment of a more complete theory of distributive justice. I shall need to specify exactly how this part is related to the whole, and I shall turn my attention to that issue in Chapters 3 and 4.

But even if the age-group question poses a distinct problem of justice, we do not necessarily perceive it distinctly. Rather, we perceive the problems raised by the aging of society as a tangle. We confuse the loose ends of distinct problems of justice with each other. We can now, however, unravel some of the strands. For example, some of the concerns about enhanced benefit programs for the elderly and reduced programs for children touch centrally on issues of justice between age groups, though the underlying question is never clearly formulated.[13] Other concerns, especially those that focus on the future solvency of Medicare and Social Security, are primarily a response to the birth-cohort question, though they may be voiced as problems about distribution between the old and the young or between generations. My point is that these distinctions are not merely semantic but signal substantively different issues about which we are often confused.

Our tendency to confuse these questions is not simply the result of collective stupidity. Rather, the confusion results primarily from our perception of competition between groups. We see the age-group question in competitive terms and it is easy to slip from that vision into the question about cohorts. It, too, is then quite naturally posed in competitive terms. I suggested earlier that the real question about age groups can be solved only if we look at the problem in a radically different way. Now I shall suggest what this "new look" involves.

Two Approaches to the Problem

People think about the problem of justice between age groups, I have argued, as one of group competition. In our aging society, the old and the young compete for scarce public welfare resources. They are locked in conflict because their group interests diverge sharply. They each want the same, limited things. The problem, then, is to determine what counts as an equitable resolution of the conflict. This competitive framework underlies many particular questions of public policy: Who shall get the life-saving transplant or dialysis if not all can—the old or the young? Is such rationing by age just or is it discriminatory? How can we make sure the old and the young are treated equally or at least with equal respect? Are we providing "generous" Social Security benefits by sacrificing the well-being of children, especially poor children?

13. Preston (1984:452) confuses the issue when he contrasts "individual" and "collective" perspectives.

Have we undermined "generational equity" if we increase per capita spending on the elderly but reduce it for children? Thus a vision of unrelenting competition dominates public policy discussion of the age-group problem.

The problem is also felt at a more personal, and not just a societal level. Here the primary perception is not one of competition, though the young feel the strain of caring for the old. Rather, there is confusion and frustration at the limits imposed by social programs. This confusion is made worse because there is deep uncertainty about moral obligations.

What do children owe their parents? Adult children—even children in their sixties and seventies—struggle to provide personal care and financial help to their aged parents. It was rare for anyone to face this burden a century ago, since so few parents lived to become the frail elderly. It is now becoming the norm. "Honor thy Father and thy Mother!" we hear and we believe. But how much and what kinds of honor are owed? Some societies have been quite explicit, tying discharge of filial obligations directly to the inheritance of property, power, and status. For us, the limits of filial obligations are less clear, especially since social obligations to the elderly are controversial and public policy is so inconsistent. We insist on extraordinary medical rescues of the incurably ill elderly, but we leave the partially disabled without services that make their lives worth living. What obligations fall to children when society does not meet its obligations to the elderly? Should society define and enforce filial obligations? (I shall return to these issues in Chapters 2 and 6.)

These perceptions of the age-group problem at both the personal and social levels share a common framework. At both levels, the issue is one of determining which transfers of goods should take place between distinct *groups* of individuals, the young and the old, adult children and their elderly parents. Which transfers are just or fair? What does one group owe the other? The sharp competition felt at the social level is tempered, to be sure, by love and a desire to care at the personal level. But the problem is still seen as one between "us" and "them." Which transfers are required by traditional notions of filial obligation? How can we design institutions that treat both groups equally, or fairly, or with equal respect?

This way of construing the age-group problem is reinforced by our tendency to see public policy as something that operates in the here and now. We scrutinize its effects at a particular time. We rarely think of public policies as instruments that operate longitudinally, over long periods of time—indeed over our whole lifespan. We ignore their impact at the various stages of our lives. When we do think about the long run, for example, when we think about what the Social Security system will be like in the next century, we shift questions. We stop thinking about the age-group question and we substitute the question about equity between birth cohorts.

This way of perceiving the age-group problem, as a problem of competition between groups viewed in a slice of time, is, I believe, fundamentally misleading. My central task in this book is to explore an alternative, the lifespan approach. Justice between age groups, I shall argue, is a problem best solved if we stop thinking of the old and the young as distinct groups. We age. The young become the old. As we age we pass through institutions that affect our well-being at each stage of life, from infancy to very old age. The lifespan approach is based on the suggestion that we must replace the problem of finding a just distribution between "us" and "them"—between groups—with the problem of finding a prudent allocation of resources for each stage of our lives. On the lifespan approach, each stage of life stands as a proxy for an age group. To determine what is fair or just between age groups we must find out when institutions treat each stage of life prudently. Clarifying and defending this suggestion will be the central task of Chapter 3. But even as a bare suggestion, the lifespan approach imposes an order on some of the perceptions of conflict we have surveyed.

Under conditions of moderate scarcity, what is made available to enhance well-being at one stage of life may not be available at others. An important task of some institutions is to help us transfer resources from a stage of life in which they are not needed to one in which they are. These institutions help us *save*. If they are prudently designed, they will transfer the right amounts of the right things to each stage of life. They will be a prudent savings scheme.

Once we transform the age-group problem into the problem of prudently designing such savings institutions, three questions become prominent. First, what is a prudent *rate* of savings from one stage of life to another? This question will be our way of approaching the problem of whether income support and health care transfers from the young to the old are at adequate levels. The problem of rationing resources by age, discussed in Chapter 5, is a special instance of this problem. Second, what *types* of transfers should we make from one stage of life to another? This question will help us address the widespread concern that our health-care system offers the wrong kinds of care to the elderly. The discussion of long-term care in Chapter 6 addresses this issue. These two questions are at the heart of the age-group problem, but they have a direct bearing on the birth-cohort question as well. A prudent institutional design must also be stable over time; it must function prudently despite the succession of distinct cohorts that pass through it. So our third basic question is: How can prudent rates and types of transfers be secured *over time?* This question is our way of approaching the birth-cohort question—the question underlying many concerns about "generational equity," as we have seen. I take up this question in Chapter 7, primarily through a discussion of Social Security. The claim I made earlier about the primacy of the age-group question means that we should answer the first two questions before trying to answer the third.

Our first order of business is to break our habit of posing the age-group problem as one of competition between groups for resources distributed in the here and now. Only by shedding this obsession with competition can we embrace the perspective embedded in the lifespan approach. But there is a second obstacle to solving the age-group problem, namely, a dispute about the relative importance of social and family obligations in meeting the needs of the elderly. Distributions between age groups have traditionally been seen primarily as a family matter, something government had best stay out of. During this century, that traditional view has been eroded. Much of the burden of guaranteeing the elderly a minimum level of income support has been transferred from individuals and families to a social, publicly administered insurance program. Similarly, the burden of providing acute health care and much institutionalized long-term care to the elderly has also been moved outside families into the public domain. In this sense, we have seen the scope of justice expanded and the role of individual or family obligations diminished, though in the United States we still leave much of the direct responsibility for long-term care to families.

Some see this "socialization" of transfers between the young and the old as a threat to traditional family values. They hanker for a return to an earlier morality. As they see it, "privatizing" responsibility for care of the elderly would revive the traditional values, such as filial obligations. Moreover, by reducing public budgets, privatizing responsibility would reduce the arena of competition between the old and the young. We would push the problem back onto the family table.

As I suggested earlier, appealing to family responsibility will not eliminate competition between age groups. This appeal will only locate the competition in the private domain, distributing its burden to a narrower group of individuals. Actually, the appeal to filial obligations only confuses us. It assumes we have a clear and useful sense of what such obligations involve and that they are an appropriate basis for public policy. I believe these assumptions are wrong, and that the appeal to individual and family obligations cannot substitute for solving the problem of justice between age groups. Against a background in which we understand what justice requires between age groups, we may be able to clarify the content of filial obligations in an aging society. But determining what is just is the primary problem; it provides a framework within which we can then solve the problem of individual or family responsibility.

This claim that the problem of justice between age groups is primary, like my earlier claim about its primacy over the problem of equity between birth cohorts, is not a matter of definition or stipulation. It is a substantive moral point that needs defense. In Chapter 2 I provide that defense.

2

Filial Obligations and Justice

The Appeal of Tradition

What is a just distribution of social goods between the old and the young? I have argued that this question poses a distinct problem of justice. The problem seems novel because the rapid aging of society has forced us to consider it explicitly, but it is actually not new. All societies, past and present, transfer income, wealth, and power between age groups. The justice of such transfers is always open to discussion. Indeed, traditional arrangements might be thought of as familiar, time-tested solutions to the problem. Appealing directly to certain traditional moral notions in order to solve the problem is therefore tempting.

In many earlier societies, transfers of social goods between age groups were primarily a family matter. Some people believe that, wherever possible, we should continue to treat the issue that way. Filial obligations, for example, should determine what the young owe the old. These obligations tell us what children owe their parents. This appeal to family responsibility is an attempt to turn back the clock. The trend in this century has been to make transfers of basic income support and many health-care services a social, not a family task.

We cannot turn back the clock. Specifically, we cannot solve the age-group problem by appealing to filial obligations. For such obligations to provide a determinate basis for social policy, we would have to have know just what burdens they impose. We cannot, however, just appeal to traditional family values and practices to determine the content of these obligations today. Neither can we find well-established moral foundations, reference to which would clarify the content of these obligations. We live in a society in which there are diverse beliefs about family responsibility, and we have neither a homogeneous tradition nor a compelling philosophical account that can overcome this diversity.

As a result, these family obligations provide a poor foundation for public policy and should not be enforced by legal sanctions. Instead, we need a solution to the problem of justice between age groups that respects the diversity. Consequently, we must first clarify our social obligations and design institutions that ensure justice between age groups. Within that framework, people can pursue their family responsibilities as their moral convictions permit. Thus it is a mistake to think we can meet social obligations by enforcing some set of individual and family obligations. We must resist this effort to "privatize" the problem of social justice.

Honor Thy Father and Thy Mother: Legislating Family Responsibility

Only one of the Ten Commandments offers a reward. "Honor thy father and thy mother: that thy days may be long upon the land which the Lord thy God giveth thee" (Exodus 20:12). When the federal or state governments appeal to traditional religious and family values, however, we should look to the costs and not the benefits. In the early 1980s, such appeals were frequent. In Hawaii, a resolution sought to reinstate a relative responsibility law that had been repealed in 1965. The bill was an attempt to correct a $30 million shortfall in Medicaid. In Massachusetts, legislation was proposed that would make each "responsible relative" of a Medicaid recipient residing in a nursing home accountable for up to 25 percent of the amount paid, though the total collected from all responsible relatives must not exceed 100 percent of Medicaid payments. According to the bill, responsible relatives are spouses, natural and adoptive children, or natural or adoptive parents of blind or disabled children under eighteen. In Colorado, legislation was proposed to require family members of a recipient of long-term care financed by Medicaid to reimburse the state for part or all of the costs, using a sliding scale of fees based on the family's ability to pay.[1]

Family responsibility laws are not new. They accompanied the Elizabethan Poor Law of 1601 (Callahan 1985:32–33), which restricted responsibility to "natural" children (Laslett 1976:94). They exist in various forms in half of the states in the United States today (Kapp 1978:287, 305). The new family responsibility initiatives do not explicitly appeal to claims about filial or other moral obligations. Instead, one can read them as simply defining a specific legal obligation for family members, without any underlying assumption about moral obligations. This would be similar to a town passing a regulation requiring next-door neighbors to water the lawns of those who go away on vacation. The ordinance might not imply or enforce moral duties to be neighborly at all. Rather, the regulation might just be an attempt to protect a common good, the appearance of the neighborhood. By analogy, we might hold families legally responsible for the care of frail elderly parents without appealing to moral obligations at all. When I argue, in what follows, that it would be unjust to legislate family responsibility by appealing to filial obligations, I am not therefore removing all possible grounds for such legislation. Still, I think it is the popular belief that people do have filial obligations requiring them to care for elderly parents that makes the family responsibility legislation seem at all justifiable.

The Limits of Filial Obligation

Duties and obligations generally impose specifiable burdens. They have limits that allow us to say when they have been discharged. Filial obligations should be no exception. If we admit that children have them, we must in general terms be able to say when they are met. We must be able to specify what it is that children owe their parents.

Traditionalists, as I shall refer to them, believe we can look to the past to specify these limits. "Traditionally," they argue, "filial obliga-

---

1. State House Notes (1984:42:7) cites Hawaii bill HR No. 54, Massachusetts bill H 1753, and Colorado House bill 1319 as examples of recent family responsibility initiatives.

tions required adult children to care for their frail, elderly parents. Children have the same obligations today, and we can specify what children owe their parents by extrapolating from traditional practice.'' Despite its recent popularity, this appeal to tradition does not help us determine the contemporary limits of filial obligations. We cannot make the straightforward extrapolation it requires for two reasons. There is a mismatch between traditional and current needs for care of the elderly, and there is a mismatch between traditional and current possibilities for care. Let's consider the first mismatch.

Since the Traditionalist appeal to filial obligations is an effort to return us to family arrangments of an earlier period, we need to look at what obligations children might have had to care for frail parents a century ago. Some demographic data is necessary. In 1900, only 63 percent of women surviving to childbearing years could expect to reach a sixtieth birthday. Today, 88 percent of such women will celebrate that birthday. The percentage of the population age 60 or over has more than doubled in the last 75 years. In 1910, there were almost 3 women aged 35–44 for each widow or divorcee aged 55 or older, but by 1973 there were only 1.2 such "daughters" for each such "mother." Finally, in 1900 only 4 percent of those 65 years and over were age 85 or older; by 1975 the percentage had doubled, and those over 85 make up the fastest growing age group in the country.[2]

What this means is that a child's obligations to care for a frail parent were far less likely to be called upon at the turn of the century, since so few people lived to be the frail elderly. These obligations were also likely to be burdensome over a much shorter period of time, since expected lifespan for the elderly was shorter. What burdens were imposed were more likely to be shared by a greater number of children per aged parent. From demographic considerations alone, it is clear that the need for care by frail, elderly parents in the past is not strictly comparable to the much greater need today. Consequently, the burdens filial obligations could have imposed in the past are not really comparable to the burdens they would impose today.

Changing social patterns, and not simply demographic changes, further complicate the task the Traditionalist faces, for they produce the second mismatch. Any specification of our actual duties or obligations requires a grasp of what it is possible for people to do. If we know

2. Treas (1977:486–487) cites the sources for the above statistics.

people cannot possibly do something, we cannot impose the obligation on them to do it. This is the point behind the oft- (and over-)stated dictum that "ought" implies "can."[3] Accordingly, to specify in a reasonable way what filial obligations require, we have to know what kind of care is possible for contemporary families to provide for their elderly parents.

Though 80 percent of all long-term care for the partially disabled elderly is now provided by families (Frankfather et al. 1981:6), when compared to the larger, less mobile families of a century ago, families today and in the predictable future may have more limited human resources with which to provide such care. Fewer adult women are at home. Delayed childrearing and the delayed entry of children into the labor market greatly extend the period during which obligations to children continue and conflict with obligations to parents. Increased mobility in an expanded economy has lead to greater dispersion of families. Moreover, many families today include complex parent and step-parent relationships that result from the increased frequency of divorce and separation (cf. Treas 1977). These changes profoundly alter the childrearing patterns that are central to the Traditionalist's claims about family responsibility. Even if it were true, as the Traditionalist supposes, that filial obligations required extensive care for parents in the past, we could not allow the Traditionalist to draw conclusions for the present directly from the past. Contemporary families are affected by altered social conditions that are not within individual control in any obvious way. For many, it is simply impossible to deliver the care we traditionally might have said they were obliged to provide. The mismatch between past and presnt social patterns changes what it is possible for families to do. This in turn changes what it makes sense to say children are obliged to do for their parents.

I have challenged the Traditionalist's belief that we can simply extrapolate current limits to filial obligations from past ones. But there is an even deeper problem with this approach. The Traditionalist assumes that in our shared past there was a Golden Age when family responsibilities were well-defined and widely respected, and his mission is to return us to it. In effect, he accepts what has been the domi-

---

3. Of course, sometimes when it becomes impossible to fulfill an obligation we undertook, for example, a contractual obligation, we are not freed of all obligation. We may be obliged to "make good" or compensate in some way.

nant view, that there is a "before" and an "after" in the treatment of the elderly (cf. Laslett 1976). That is, *before* industrialization or modernization, the elderly received respect and care and were generally incorporated into multigeneration extended families. *After* modernization, the elderly receive little respect or care, and households have generally been restricted to the nuclear family of parents and nonadult children. This dominant view is largely mythical (cf. Laslett 1972, 1976; Gordon 1978). Many sociologists counter the myth of the "after" by documenting the extensive care provided to the elderly by today's modern family (cf. Shanas 1979; Shanas et al. 1968; Frankfather et al. 1981). (I shall return to this point in Chapter 6.) Here it is more important to see what is mythical about the "before" part of the dominant view, the belief in the Golden Age of the family.

One line of carefully compiled evidence about the "before" is based on studies of pre-industrial England, which has some claim to being a central part of "our" past. Less than 10 percent of households had more than two generations in them, and levels of care within families were similar to those found today. There was no recognized social duty to give care or family membership to aged relatives other than the parents of the head of a household (Laslett 1976:93–94). Though frail fathers and mothers were sometimes brought into the households of married children, it was not a widespread occurrence and by no means a universal pattern. Moreover, where solitary or needy parents or other relatives were brought into the family, it seems to have been because of the benefits brought by the help they could give with childrearing. There is little evidence the patriarch had the sanction necessary to require a child to live with his parents and provide care (Laslett 1976:95). The old as well as the widowed tended to live in their families or households as before the advent of frailty or widowhood. They were not absorbed into other households as the myth of the "before" would have it. The Traditionalist would be hard pressed to find his Golden Age actually existing in the English tradition.

Some other cultures have elements that resemble the "before" more than the English. For example, in Estonia in the seventeenth century, in Latvia in the eighteenth century, and in Hungary in the early nineteenth century, many more widowers lived in families with married children than in England (Laslett 1976:114–115). In many cases where we find children undertaking care of their elderly parents, however, the obligation is not what the Traditionalist seemed to have in mind. One child might acquire special obligations to stay with the father and care for him, but the care provided was commonly given in exchange for special property privileges (cf. Goody 1976:118; Goody 1958). In many cultures, when property was allocated to a son or daughter at the time of marriage, the parent reserved a right of support from the child. Reserving these rights did not always guarantee adequate care. In some pastoral societies, after divestment of flocks by elderly parents, subsistence of the elderly was often quite marginal (see Stenning 1958, cited in Goody 1976:120). In Scandinavia, an arrangement known as *Flaetfoering* existed. A parent who divided his property among his children was then entitled, by law, to make a circuit of their households, spending time with each of them in proportion to the share each received (Goody 1976:121).

What the Traditionalist nostalgically sees as a warm network of love and family responsibility appears to the colder historical eye as a complex mechanism in which care was exchanged for access to the means of production (albeit, no doubt at least some of the time with a sense of love or moral obligation). Variations in the mechanism often reflected differences in the kinds of property transmitted and the types of control over future livelihood the parent could exercise. My point here is not to support or to appeal to the strong Marxist claim that morality is just ideology. Nor am I suggesting that there is no validity to the notion of filial obligations because such obligations arise in order to stabilize certain rights to property or relations to production (cf. Cohen 1978). I am committed to a far weaker claim: Even if there is validity to such moral notions as filial obligations, we cannot simply abstract past patterns of family responsibility from the economic and political context in which they functioned and then insist that these patterns "ought" to apply within our vastly different institutions and social practices.

The problem faced by the Traditionalist, who wants to resurrect the Golden Age of family responsibility, is that there is no *one* "before." There are *many*. Even if he could solve the problem of how to extrapolate from the past to the present, we do not all share a single past. The English tradition does not match the "before" of the Traditionalist's vision. Laslett (1976:95) puts the point quite humorously:

> The ordinary story of the [English] family-household after the childrearing stage was of offspring leaving successively, though not necessarily in order of age, until the parents finally found themselves alone if

they survived. There is a telling contrast here with the traditional familial system of an area like South China, for example, where no child left the parental household except under clearly specified conditions because the recognized rules of familial behavior required coresidence wherever possible, and where no father or mother of grown offspring would ever live alone. Some part at least of what has been called "the world we have lost syndrome" must be attributed to English and English-speaking peoples being led to believe that their familial past has been the same as that of people like the Chinese.

Furthermore, important features of other traditions, which more closely resemble elements the Traditionalist wants to resurrect, seem rooted in property relations and patterns of life to which we could not return no matter how strong our nostalgia.

I shall return to this point about the diversity of patterns and traditions of filial obligations later in this chapter, since it has a bearing on the degree to which we should appeal to them as a legal basis for social policy. Here I want to emphasize a different implication of my argument. The Traditionalist approach to specifying the limits of family responsibilities will not work, I have argued, because past needs do not match modern ones and because new patterns of family life make providing some kinds of care impossible. More important, the central element in the Traditionalist vision, the appeal to a Golden Age of family responsibilty, is an appeal to a myth rather than a social history we share. Therefore we cannot justify introducing a social policy that enforces some set of family responsibilities on the grounds that it merely sanctions well-understood, traditionally recognized obligations. Such legislation would not be appealing to *prior* obligations that we all embrace as a result of sharing the same tradition. Rather, it would amount to a *new* definition of these obligations and their limits, one that needs a justification that goes well beyond an appeal to tradition.

Foundations for Filial Obligations

We would be in a better position to determine the burdens imposed by filial obligations and the limits appropriate to them if we had a clearer grasp of their moral foundations. What are the reasons for recognizing such obligations? Why should we honor our fathers and our mothers?

Setting aside appeal to divine authority (the Ten Commandments) or to cultural tradition (the mythical Golden Age of healthy family values), what philosophical account can we give?

One common view is that these obligations are the corollaries of parental duties to care for children. A child is wronged by his parents if adequate care is not given him, and the parent violates a duty if he or she neglects to give such care. This duty of the parent is quite general and may be satisfied by doing a wide variety of things for the child. Corresponding to this duty of the parents to provide care are rights of the child to receive it. But it is not this correspondence that concerns us here. Rather, the view in question is that corresponding to the parental duty there is a filial duty or obligation to respect and obey the parent. A special instance of this corollary duty is the duty to care for the parent when the parent needs or requests it.

This common view ignores a basic asymmetry between parental and filial obligations. Some duties, such as the duty not to harm others or the duty of beneficence, may be thought of as *natural* duties, duties which we have by virtue of being rational, social creatures. But most duties arise because of particular roles we assume or agreements we make. Parental duties to children are usually thought of in this way. Parents assume the duties of caring for their children through their own adult acts. The duties that come with the role of parenting are ones that parents undertake. They bring their children into existence—or they adopt them—and it is this act that imposes duties on the parent.

Are filial duties in any similar way self-imposed? What acts of children could be the basis for the duties which this common view attributes to them? Children did not ask to be brought into existence. Moreover, their desire for care and their need for it, once born or adopted, cannot be the sole basis for claiming they have "implicitly consented" to being bound by the duty to care for their parents. The role of being a child is not one we undertake in the way we undertake the role of parenting. For filial obligations to derive merely from our having had the role of being children, we would have to explain why some duties arise only out of roles we assume, whereas others arise out of roles that are inevitable or unavoidable or that can be imposed on us by others. That is, we need to know how the self-imposed duties of parents can give rise to the non-self-imposed duties of children. Without this further account we cannot accept the view that children's obligations are a mere corollary of parental duties. The asymmetry between parents, who deliberately adopt their roles, and children, who do not, points to a real gulf in the grounds for the corresponding duties.

Despite this asymmetry, the common view might be defended in another way. Without the respect, or at least the obedience of children, parents may not be able to discharge their duties to provide adequate care for children. Similarly, a teacher has a duty to instruct her students, but she cannot succeed unless the students are obedient and follow her instructions. The obedience of children and students to parents and teachers seems to be a necessary condition for parents and teachers to discharge their duties successfully. But even if this were to persuade us that the duty of children to obey parents is a corollary of the duty of parents to provide them with care, it does not help us prove that *adult* children have filial obligations to care for their elderly parents. Students need not obey teachers after they graduate, and adult children are not in general obliged to obey their parents in most matters. Similarly, however much students must obey teachers, they do not have to educate their teachers after they graduate. This defense of the common view proves too little.

There is another way to try to explain the view that filial obligations are corollaries of parental duties. On this view, parental duties give rise to corollary filial obligations because of the good things that parental duties deliver when they are discharged. The parent who does his duty delivers a benefit to his child, and the beneficiary of such a duty acquires an obligation to reciprocate.

This view is not plausible if it rests on the principle that, whenever good things are done for us out of duty, we are obliged to reciprocate. If I am a paramedic and fulfill my obligations by resuscitating someone, he is not thereby under similar obligations to rescue me, however grateful he may feel. If a crossing-guard halts traffic so that my son can get safely to school, my son has no duty to reciprocate. If I discharge my duty to contribute to charity by giving to Oxfam, the starving villager in the Sahal is under no duty to return some good to me, whatever his gratitude. In general, if someone has a duty or obligation to provide me with some good, I do not thereby incur an obligation to return the good or its equivalent.[4]

There are two ways to save the view that the good done for children by parents gives rise to a duty to reciprocate. One way is to explain why

4. For similar points see English (1979) and Callahan (1985).

parental duties, unlike other duties, imply reciprocal duties. But proving that parental duties are exceptional in this way seems an unpromising strategy. We cannot simply declare them different without abandoning the search for a principled, philosophical account. The alternative is to abandon the belief that parents have a duty to provide the good they do for their children and to invoke instead a very general principle of reciprocity. This principle requires people to reciprocate for good acts done on their behalf whenever the good is not owed to them, for example, because it is only a favor. This principle of reciprocity in effect says it is only fair to return the favor of good things done to one in the past, especially where they might have involved considerable sacrifice, in this case by parents. In this way, children fall under an obligation to reciprocate for the good done them by parents. But then, so do a lot of other people fall under similar obligations to those who bestow goods on them.

This appeal to a general principle of reciprocity will not work. One obvious difficulty is that we can no longer speak of parental *duties* to provide for their children: We are obliged to reciprocate good favors done for us, not goods owed us. A second difficulty is that a principle of reciprocity must take the good with the bad. Some children receive harms and not just benefits from their parents. If filial obligations rest on reciprocity, then we have to examine quite carefully just what individual children owe their parents. Some children might owe their parents nothing, and some might owe them harm. Since children and their parents might very well assess the goods and harms differently, we would have on our hands a complex matter of determining specific debts, not a general presumption that reciprocal benefits are owed.

The appeal to reciprocity presupposes that it is "favors" that parents have given children. We do owe a return of favors done for us. Even parents will respond to children's requests for favors by specifying what children have to do to earn them or to return them. "Please take me to the movies!" may be responded to with: "Only if you clean up your room." But as Jane English (1979:352) has argued, much that parents do for children is done because parents unconditionally want good things for their children. They love them and have expectations about how they want their children to grow up. Doing good things for children is one of a parent's projects in life that makes life meaningful—for the parent. Such actions are not mere favors and do not give rise to debts on the part of children. Indeed, society may be invading a quite intimate relationship, in which unconditional love and not conditional favors are dominant, if it tries to coerce children into repaying their parents for good things done out of love (cf. Shoeman 1980).

English (1979:351) argues that what adult children *ought to do* for their parents grows out of bonds of friendship they feel for their parents, not out of debts they owe for past favors. There are problems with this view and the analogy on which it rests. The friendship that exists between adult children and parents is quite different—in intimacy, origin, and complexity—from typical adult friendships, for example (cf. Callahan 1985:34). Moreover, the appeal to friendship does not account for the belief, reported by some adult children, that they owe their elderly parents some form of help, even if they want no "friendly relations" with them. One may be able to dismiss these feelings of obligation as the result of confusion or uncritical acceptance of custom, thus undermining this appeal to filial obligations. Alternatively, we might try to save this intuition by looking in another direction for a philosophical defense of these obligations.

One such direction is the claim that filial obligations are really only a special case of the (imperfect) duty of beneficence. In general, the duty of beneficence allows us to pick and choose where, when, and to whom we will do beneficial things. In some circumstances, however, we may be under special obligations to help particular people because of their acute need and our unique ability to help. On this view, children have obligations because they are in a unique position to help parents. It is the view someone might express as follows: "It's not that I owe it to Mom for what she's done, but there's no one else she can turn to and it would be wrong to turn my back on her need."

This view explicitly abandons whatever might have seemed plausible about the view that children owe parents something because of what their parents have done for them. Instead, on this view, children must care for their parents only because they are well-situated to help. The view also inherits the difficulty that attends all attempts to enforce mere duties of beneficence as if they were perfect duties. We usually allow people to give to the charity of their choice; we do not require that such good deeds be done for specific people.[5] Finally, duties of beneficence

5. Allan Buchanan (1984) has argued that the enforcement of some duties of charity may be justified because coordination problems otherwise block effective charitable action. Whatever the merits of his argument in the case of taxing people to provide health care for the poor, one cannot easily appeal to it here.

do not require us to perform very demanding and burdensome deeds. Such performances are generally viewed as supererogatory, not obligatory. Where persons are particularly well-situated to deliver a highly beneficial good or prevent a serious harm, as in the case of the good swimmer who alone is close enough to rescue safely a drowning victim, we may think the duty of beneficence becomes more restrictive or more perfect. Some children, we may concede, are especially well-placed to deliver long-term care to their parents. But providing long-term care for disabled parents is exceedingly burdensome, unlike the rescue by the good swimmer. It is just the sort of thing that would be particularly difficult to justify as a requirement of charity. A very burdensome or dangerous rescue is generally praised as an act above and beyond the call of duty.

There are other philosophical strategies that may yield a basis for filial obligations. For example, caring for parents, we may believe, is a virtue that ought to be encouraged. We are better people for having it. Perhaps this virtue is part of some moral ideal. Still, this strategy requires that we must not only defend this ideal as morally praiseworthy, but show that achieving it is a moral duty. A more promising strategy, which I develop further in Chapter 6, is to argue that certain socially important, traditional relationships that provide fundamental goods to individuals, such as those occurring within families, may give rise to special duties. The central difficulty facing this approach is to explain why some traditional relationships give rise to such duties but others do not. In any case (as I argue in Chapter 6), special duties defended in this way require that a general background of just arrangements exist. Filial obligations, defended in this way, require that society meet its social obligations to the frail elderly. Because they presuppose social obligations, filial obligations cannot be the basis for defining social obligations.

This survey of possible grounds for filial obligations does not prove they are all losers. My intention has been only to show that there is no obvious winner which we are compelled to accept and to use in specifying the limits of such obligations. The view that these obligations are direct corollaries of parental duties might have given us the most direct determination of what children owe elderly parents. Since we already provide legal sanctions that enforce parental duties, extending such sanctions to filial obligations might have been defensible, were they real corollaries. But this account also faced the most serious objections.

Looking at filial obligations as mere reciprocation for the good favors we received as children yields a much less specific—and highly individual—determination of what children owe their parents. It also abandons the common and reasonable view that parents have duties to children. The view that the obligations children have to parents are just special cases of the duty of beneficence makes it implausible to enforce them through legal sanctions. In any case, duties of beneficence can require us to undertake only modest burdens.

We remain without compelling foundations for filial obligations, even though it still may strike us as rational and fitting and praiseworthy that (most) children want to help their parents. Without such foundations, however, we cannot specify the content and limits of filial obligations—and it is clear, given today's demographics and social structure, that there must be limits if such obligations exist at all. Without agreement on these limits, filial obligations cannot be made the basis for laws enforcing family responsibility.

Justice Despite Diversity

Social policy concerning the young and the old should rest on a solution to the problem of justice between age groups. That solution should count as a court of final appeal; it should give us a public basis for resolving disputes about policy. Some people believe that a clear understanding of what children owe their parents would give us that solution: We could then make family responsibility the basis of social policy. Unfortunately for this strategy, filial obligations remain a singularly indeterminate solution to the problem. The Traditionalist approach failed because it could find no shared heritage from which we could unequivocally extrapolate the limits of these obligations. Similarly, we have not been successful in finding well-established moral foundations for them, and thus we have no clear grasp of their limits. Both Traditionalist and philsophical approaches failed to provide us with a publicly understandable basis for producing agreement about the content of these obligations. This failure should make us leery of attempts to base social policy on appeals to these individual obligations.

The problem is not, of course, that individuals have no idea what filial obligations entail. In fact, most people have quite strong, deeply held moral convictions. Some vehemently deny that they have any such obligations at all. Most insist just as firmly that they have them, but they may disagree about what the obligations entail. Those who believe they owe their parents extensive care may insist that others, who do not so believe, are immoral shirkers of duty. Those who do not believe they have such obligations will resent believers trying to impose obligations through legal sanctions. The diversity of cultural traditions, which frustrated the Traditionalist, may explain some of this diversity in current beliefs. Similarly, the absence of well-established moral foundations for filial obligations also explains the variety of views. Whatever the explanation, this diversity is a fact of our social life and not likely to disappear.

Of course, society could make these obligations determinate by fiat, by passing laws governing family responsibility. The difficulty with doing it that way is that social policy needs to be *just* or fair as well as determinate. The obligations we specify—even by fiat—should result in a fair system of transfers between age groups. Not merely any determinate set of legal obligations to parents will do. Moreover, the rationale we offer for the transfers we consider to be fair must be one we can publicly defend. The problem we have uncovered is that we lack any persuasive way to sanction by law one uniform set of filial obligations given the disagreement in moral beliefs about them.

A fundamental point about the nature of justice is at the heart of this problem. Principles of justice must yield a framework of institutions within which people having different views about what is good and right in other regards can cooperate. In general, and especially in heterogeneous societies such as ours, individuals will have strikingly varied conceptions of how to live a good life. They will have different fundamental goals and projects in life, so their lives will achieve value or meaning in a variety of ways. We might think of the "plan of life" that connects these goals, projects, and preferences into a rational pattern of activity as defining "the good" for an individual. (The term "plan of life" derives from Mill and is given prominence in Rawls 1971.) At least some aspects of individual morality, that is, some beliefs about what is right for us to do, will also vary, and these variations form part of the diversity in plans of life. Within our society, the cultural variation in beliefs about what children owe parents helps shape individual conceptions of the good life in different ways. But it is the task of a theory of justice to provide us with principles that can act as a final and publicly acceptable basis for resolving disputes about how basic social goods, such as liberties, opportunity, income and wealth should be distributed among people who disagree about many other things. Justice must provide a framework for cooperation among individuals who may nevertheless disagree about much of what is good in life and about many of the ways they treat each other as individuals.[6]

In view of the aging of society, we face both a moral and practical problem. We must justify a system of transfers of social goods between age groups that accommodates new facts about our demography. As a result, we must consider explicitly a problem of justice that has often been left to tradition. We must discover and be able to justify principles that will govern social policy concerning the young and the old, and we must do so despite the fact that people disagree about what children owe parents. This means that we need a perspective from which we can arrive at a solution to the age-group problem without having to resolve these disputes about family responsibility. We need to solve the problem of justice in a way that allows people to pursue their own moral beliefs about family responsibility. We must define what our social obligations are in the distribution of goods between age groups, and we must design institutions that meet those obligations. At the same time, these institutions must allow diversity in the pursuit of family relationships and family responsibility. The problem of social justice is primary here: We must solve it collectively whatever other variability we allow individuals in their conceptions of what their individual or family obligations are.

This general point about the relationship between justice and family responsibility does not mean that social policy should ignore the responsibilities many children feel for their parents. Even if we abandon all talk about filial *obligations* (and I do not), we need not ignore the fact that many adult children care deeply about what happens to their parents. They believe they *owe* their parents significant efforts to care for them, and they do in fact provide such care. Sociologists have, as I noted, debunked that other myth about families, namely, that the elderly are not being cared for by their children. Enormous efforts are being made by extended, multigeneration families to provide long-term care for their elderly (cf. Shanas 1979). Moreover, this is usually a type, quantity, and quality of care that the public sector is unlikely ever

6. This picture of the role of justice is developed by Rawls (1971, 1980, 1985).

to provide a substitute for. Were this supply of services to disappear or to be undermined, it would be an unmitigated disaster for the elderly. My point throughout this discussion is that it is morally wrong to protect the supply of family care by legally enforcing some set of filial obligations for which we can provide no adequate moral justification. Rather, a just health-care system should meet social obligations concerning care of the frail elderly and, at the same time, be responsive to the importance and fragility of the family's contribution to long-term care. It should include institutions designed to encourage such family care. I return to further discussion of this issue in Chapter 6.

"Privatizing" Versus Social Justice

I have portrayed the appeal to family responsibility on the part of the Traditionalist as a nostalgic effort to turn back the clock (albeit to a mythical past). In trying to solve the social problem of justice between the young and the old by appealing to individual moral obligations, in "privatizing" the problem, the Traditionalist seeks to reestablish a more robust set of "old style" family values. What the young owe the old becomes the problem of what children traditionally owe their parents. The suggestion is that we can achieve social justice through individual morality, provided we enforce it legally.

The yearning for traditional individual values actually feeds a broader movement to "privatize" the problem of justice between age groups, indeed, to "privatize" many other issues of social justice as well. For example, some critics of the Social Security system would like to replace our socially administered system of transfers between age groups with a largely private system of individual savings and annuity plans. Similarly, wherever possible, these critics want to transform the social insurance plan called Medicare into one that relies more directly on individual ability to pay for care, for example, through cost-sharing. Specific reforms in the direction of "privatizing" social transfers are defended on various grounds, but this broader movement has its roots in certain libertarian views about justice itself. It is important to see why the Traditionalist and the libertarian unite in their support for "privatizing" transfers of services to the elderly.

The libertarian is interested in "privatizing" because of basic features of this theory of justice. With the libertarian view (cf. Nozick 1974), distributions of resources are just if they have the proper pedigree or history. They must arise through free exchanges of goods by individuals who are entitled to the property they seek to exchange. Such entitlements to goods exist where individuals acquired their property "from nature" in a just fashion or through exchanges with other individuals who themselves were entitled to it. In its pure form, this theory stands opposed to redistributions of property by the state, except to rectify past injustices. The pure libertarian is opposed to a state that redistributes property even to maximize the liberty of its citizens. He is also opposed to any property transfers that are undertaken to enhance the opportunity or welfare of citizens, even if some may think such transfers make the system more fair. Justice, in this view, arises through individual actions which are fair because they are freely made. Social justice is the outcome of individual, not societal activities. The libertarian opposes, then, the trend, dominant in this century, of enhancing welfare, including the welfare of the elderly, through redistributive transfers of goods.

No social obligations exist for the libertarian, except to defend the society against external interference and to protect individuals from violations of their rights to property and security. Individual morality therefore assumes considerable importance in protecting individuals from the roughness of life and in providing a sense of community and mutual responsibility. The libertarian is likely to emphasize the importance of individual charity and other individual moral virtues that preserve some fabric of interpersonal responsibility.[7] In this spirit the libertarian joins the Traditionalist in appealing to filial and other individual obligations that enhance family responsibility.

This wedding is one of convenience, however, and not necessarily one of shared commitments. The libertarian has no basic commitment to save traditional family patterns or values from eroding in the stress of modern life. In turn, the Traditionalist may not have any fundamental

7. Allan Buchanan (1984) has argued that this emphasis on individual charity may be of little consolation to those who are really worried about the needy. Individual charity tends to be inefficient for many important kinds of needs, and strictly voluntary collective charitable efforts are likely to fail because of familiar obstacles to successful collective action, such as free-rider probelms, problems of assurance, and coordination problems. Because of these problems, the libertarian may have to allow the state to enforce coercively some charitable duties.

aversion to a state that redistributes goods to promote equality or fairness. Some libertarians may also be Traditionalists, but such convergence in faith is not the general case.

The main argument of this chapter, that filial obligations remain indeterminate in a way that makes them an inappropriate basis for social policy, affects the partners in this marriage differently. It is a greater threat to Traditionalists than to libertarians, who may not want the state to enforce these obligations by law in any case. Traditionalists, we may suppose, remain concerned that society promote justice between the young and the old and, therefore, will be motivated to abandon the appeal to filial obligations in favor of a socially workable solution. They can still promote family responsibility within a framework of institutions that guarantee justice between the old and the young. Thus Traditionalists should divorce themselves from the libertarian's quest for privatization. Unlike libertarians, they have no principled stake in dismantling the social institutions that provide transfers of goods between age groups. If I am right that traditional family values are more likely to survive when we provide a framework of just institutions, Traditionalists should disavow the libertarian commitment to privatization.

In what follows, I shall not try to refute the general libertarian framework, since that would be the task of a quite different book. Rather, I shall develop a solution to the problem of justice between age groups that stands as a clear alternative to what any libertarian would propose. (Of course, describing the distinct solutions to particular issues of justice, implied by different general theories of justice, can help to persuade us that one general theory is better than another.) My alternative to the libertarian approach is not, however, incompatible with the underlying concern of the Traditionalist, to promote a robust sense of family responsibility. Rather, my approach can embrace such values.

3

The Prudential
Lifespan Account

Inequality: An Ambiguity

The key to solving the problem of justice between age groups is acknowledging the humbling fact that we age. To see why this banal fact is so important, consider what I shall call the Inequality Objection. According to this Objection, there is no special or distinct problem of justice which the aging of society forces us to confront.

The central concept in distributive justice, so the Objection goes, is that of inequality. Principles of distributive justice tell us what inequalities in the distribution of important goods are morally permissible. Specifically, important principles of distributive justice prohibit our using "morally irrelevant" traits of individuals, like race, religion, or sex, as a basis for differential treatment in the distribution of important social goods, such as educational or job opportunities, civil liberties, health care, income and wealth. In distributive contexts we must consider only morally relevant facts about individuals: talents and skills in hiring and promotion, health status in medical settings, performance (including special deficits or gifts) in educational institutions, and poverty, disability, and employment status in income support programs. But just as race or sex is a morally irrelevant trait, so too is age. Indeed, we recognize this moral point in our laws against age-discrimination, which prohibit treatment that discriminates by age in employment, housing, and other contexts. What this means is that there is no special problem of justice between age groups any more than there is a special problem of justice between whites and blacks. Justice demands that we treat people equally, ignoring their morally irrelevant traits, such as race or age. We should allow unequal treatment only when there are morally relevant differences between persons.

This Inequality Objection centers on two distinct claims, one about the moral relevance of age, the other about inequalities between persons. The first claim is that, because age, like race or sex, is a morally irrelevant trait, it cannot be used in ways that promote inequalities between persons. It is not self-evident, however, which particular traits are morally relevant or irrelevant to distribution. It takes substantive moral argument to establish what is relevant in particular distributive contexts. I postpone the problem of deciding whether age is ever morally relevant until Chapter 5.

Here I reply to the Inequality Objection by rejecting its second claim, that differential treatment of people according to age, like differential treatment by race or sex, always generates inequalities between persons. The claim is clearly true for race or sex. If our institutions treat blacks and whites differently, giving certain opportunities only to blacks, others to whites, then we generate an inequality between persons. (Sometimes we may need to treat blacks and whites differently in order to rectify an unjust inequality, but the complex issues raised by a policy of affirmative action are beyond the scope of my essay.) Since race *is* a morally irrelevant trait, then the inequality generated is based on a morally irrelevant difference between persons.

Age is different. Remember the banal fact: We grow older, but we do not change our race or sex. If we treat the young one way and the old another, then over time, each person is treated both ways. The advantages (or disadvantages) of *consistent* differential treatment by age will equalize over time. An institution that treats the young and the old differently will, over time, still treat people equally. Whereas differential treatment by race and sex always generates inequalities between persons, differential treatment by age does not necessarily generate inequalities. Therefore, even if age is a morally irrelevant trait, using it in certain distributive contexts will not generate an inequality in life prospects for morally irrelevant reasons—because it generates no inequality at all. (Of course, I am assuming that the policy is stable over

time and that all people go through the whole age range. This idealization ignores "start-up" problems that arise when we begin such policies. Some of these issues are addressed in Chapter 7 when I discuss
equity between birth cohorts.)

We can see the importance of this difference if we ask the question:
Are we primarily concerned about unequal treatment at a moment or
over a lifetime? For most historically important traits, such as race,
religion, or sex, it does not matter how we answer this question. A
pattern of differential treatment by race or sex at a moment will lead to
differential treatment over a lifetime, for these are fixed traits of individuals. Where these characteristics are at issue, then, there are definite
inequalities and, therefore, problems of justice to be confronted. But a
consistent pattern of differential treatment by age, over time, will erase
the inequality it seems to entail, as long as that differential treatment is
consistently administered. Our banal fact points to an ambiguity in the
phrase "unequal treatment": Differential treatment by age, over time,
is *not* unequal treatment of persons, even if it is unequal treatment of
age groups on each occasion, at each moment.

Because the Inequality Objection does not acknowledge this ambiguity, it fails to deal convincingly with questions raised by the aging of
society. It fails to show definitively that there is no distinct problem of
justice between age groups. Our banal fact means that we cannot rule
out unequal treatment by age simply through a direct appeal to equality
in the way we can rule out unequal treatment by race or sex. The basic
question remains: *Which* unequal treatments of age groups are just or
fair?

Subsidies and Savings

Because we age, treating people of different ages differently does not
mean we are treating persons unequally. From the perspective of institutions that operate over a lifetime, unequal treatment of different age
groups does not generate inequalities among persons. Indeed, unequal
treatment at different stages of life may be exactly what we want from
such institutions. The lifespan account of justice between age groups
builds on this basic point.

As we age, we pass through social institutions responsible for distributing important social goods to us, such as income and health care, in accordance with various distributive principles. Since our needs vary at different stages of our lives, we presumably want these institutions to be responsive to these changes. It is prudent to design institutions so that our entitlements, our legitimate claims to goods and services, reflect our changing needs. Because goods and services are not available in unlimited quantities, we must be prudent about which ones we claim at one stage of our lives lest we deprive ourselves of important benefits at another. In general, budgeting prudently enables us to take from some parts of our lives in order to make our lives as a whole better. Specifically, this may take the form of "saving" resources by deferring their use from one stage of life to another.

Consider an institution that actually does help us to budget over our lifespan. In the United States, we defer income from our working lives to our post-work retirement period, which has generally been after age 65, in two main ways. We may contribute tax-sheltered income to group and individual pension plans during periods in our lives when our earnings are high, drawing on them when we retire and our income is lower. This tax policy is a social subsidy that provides incentives for us to accumulate vested assets. It reinforces our individual interest in deferring the use of resources from one stage of life to another by raising the individual cost of using resources without deferment. We are thus encouraged to balance long-term against short-term self-interest. This public subsidy of retirement plans reduces the degree to which post-retirement income support must depend on later public subsidies. Though we may think of such pension funds as private or individual transfers, the existence of a tax policy promoting them shows they have a social dimension as well. As history shows, society needs institutions that facilitate income transfers from one stage of life to another.

We also defer income from our working lives to our retirement period by paying a payroll tax. This tax revenue from the employed is used to fund the Social Security benefits of currently retired workers. In this scheme there are no *vested* savings, and the income transfer from young to old also involves some income redistribution to poorer retirees. But, if the Social Security system remains stable, young workers will be entitled to claim benefits when they age and retire. There is an intergenerational compact that has the effect of transfering resources from an individual's working years to his retirement years, insuring that basic needs can be met over the whole lifespan. Moreover, it tranfers income in a way that is relatively well-protected against inflation and the uncertainties of predicting individual lifespan (Parsons and Munro 1978), even though the system may need adjustments to allow for shifts in birth and economic growth rates (see Chapter 7).

Our health-care system has a similar "saving" function, though we do not often notice it. This saving function is more apparent when there is a comprehensive national health service or insurance program. To see it in the United States, however, we must think of our private and public insurance schemes as one set of institutions, adding Medicare and Medicaid onto our system of largely employer-funded health-care insurance. We then notice that employed workers—young and middle-aged adults—pay the overwhelming share of all health-care costs. On the average, their annual health-care insurance premiums—partly in the form of employer benefits, partly in the form of their own payments, partly in the form of payroll taxes—far exceeds their actual health-care costs. There is a transfer of health-care resources to children, and even more dramatically to the old. Indeed, those over age 65 use health-care dollars at roughly 3.5 times the rate of those under 65 (Gibson and Fisher 1979), though it is a relatively small proportion of high-cost users among the elderly who account for this statistical increase (Zook and Moore 1980). In effect, society is community-rated as one risk pool, despite great differences in the health status of different age groups. The relatively healthy adult working population pays a premium that exceeds what is actuarially fair to it. Age groups are treated differently. The old pay less and get more, the young pay more and get less.

Viewed from the perspective of an institution that operates over our lifespan, however, the "unfairness" of this inflated premium is an illusion. What is crucial about the health-care system is that we pass through it as we grow older. The system transfers resources from stages of our lives in which we have relatively little need for them into stages in which we do. We pay for health care we do not use in our middle years, but we receive health care we do not pay for in childhood and old age. We feel we pay through the nose as working adults, but we are free-loaders in youth and old age.

If our perspective is that of working adults who think only about their inputs and outputs in that stage of life—which is, understandably, our everyday perspective—then the inflated premium may seem unfair. If, however, we think of the system as a social savings scheme, our cross-subsidy to others vanishes. That is, if we consider the benefits the system gave us for health care we did not pay for in our youth, and if we think about the benefits we will be entitled to without paying (fully) for in our old age, then the picture changes. We see then that the inflated premium in *our* adult years is needed to pay for *our* needs in other stages of *our* lives. Of course, our actual insurance scheme is not prudently designed in many of its features, as I suggest in Chapters 5 and 6. Still, we all benefit from an institution that reallocates health-care resources from stages of our lives in which we have many resources and few needs into those stages in which we have fewer resources and greater needs.

Two basic points thus emerge from our discussion of the unequal treatment of different age groups. First, such unequal treatment does not mean that persons are treated unequally over their lifespan. Second, such unequal treatment may have effects which benefit everyone. These points are not hypothetical. We already try to take advantage of them in institutions and policies that allocate income support and health care over our lifespan.

These two points provide the central intuition behind the Prudential Lifespan Account of justice between age groups. The lifespan account involves a fundamental shift of perspective. We must not look at the problem as one of justice between distinct groups in competition with each other, for example, between working adults who pay high premiums and the frail elderly who consume so many services. Rather, we must see that each group *represents* a stage of our lives. We must view the prudent allocation of resources through the stages of life as our guide to justice between groups.

The suggestion that we shift perspective may seem a conjurer's trick. Now you see an inflated insurance premium, now you don't. Now you see the "others" (the elderly or the young benefiting from what is "yours"), and now you don't. But this shift of perspective is no trick. It is made reasonable by that elementary fact I pointed to earlier, that we each age, and by the two basic points that we have derived from that fact. What we need to see is ourselves at other stages of our lives, benefiting from our own (albeit imperfectly) prudent savings.

The contrast of age with race or sex is again illuminating. A health-care system that treats age groups unequally does not generate inequalities between persons over time. If prudently designed, it can benefit *each* person. A health-care system that treats blacks and whites, or women and men, unequally will generate inequalities between persons over time. It will *not* benefit each person; it will be open to the objection that it takes benefits away from some people in order to help others (and for morally irrelevant reasons).

This contrast has profound implications for distributive justice. From the perspective of stable institutions operating over time, unequal treatment of people by age is a kind of budgeting within a life. If we are concerned with net benefits within a life, we can appeal to a standard principle of individual rational choice: It is rational and prudent that a person take from one stage of his life to give to another in order to make his life as a whole better. If the transfers made by an income-support or health-care system are prudent, they improve individual well-being. Different individuals in such schemes are *each* made better off, even when the transfers involve unequal treatment of the young and the old.

We cannot, however, treat people unequally by race or sex within institutions that transfer goods and achieve the same, unproblematic results. We would be treating different *persons* unequally, benefiting some people at the expense of others—even if the transfers promoted aggregate well-being and in that sense made society as a whole, or on the average, better off. Thus taking from one race in order to benefit another, or to benefit society as a whole is morally highly problematic. Taking from some to give to others on the basis of race crosses the boundaries between persons, and distributive schemes that cross personal boundaries in this way risk failing to respect the importance of persons. We cannot simply generalize the principle of rational choice for individuals, which applies to stages of a life, and convert it into a principle of social choice, which applies to different persons.[1]

The difference between the race and age cases should by now be clear. Distributive schemes that take age into account look like cases in which we cross boundaries between persons only if we adopt the perspective of a moment or time-slice. From this perspective, the "me" of the adult worker "now" subsidizes health care for "them," the (demanding) elderly, or for "them," those (darling) children. Once we take the longitudinal perspective of institutions operating through time, the appearance that we are crossing the boundaries between persons fades. The shift in perspective I urge is thus rooted in a real, distinctive,

---

1. This point is the core of Rawls's (1971, 1982) criticism of utilitarianism; cf. Daniels (1979) and Parfit (1973, 1984), and the Appendix to this essay.

and morally important fact about age. It is no philosophical sleight-of-hand.

"Framing" the Age-group Problem

To find out what is just between age groups, we must seek principles to govern the design of the institutions that distribute goods to us over our lifespan. From the perspective of such institutions, transfers between age groups appear as transfers between the stages of a life, not between persons. The shift in perspective I have been urging thus means we should not seek the typical principles of justice which govern distribution between competing individuals or groups. Rather, we must seek principles governing allocations within a life. The fact that this shift is plausible in the case of age groups is what makes it a distinct problem of distributive justice. For it, but not for problems of distributive justice in general, prudence is a safe guide to justice.

How should the appeal to prudential reasoning be made in this kind of case? Two distinct issues must be faced. First, we must be clear about the *scope* of the age-group problem and about the appeal to prudence. The relevance of prudential reasoning to its solution is what distinguishes the age-group problem from more general problems of distributive justice. Because it is distinctive in this way, we cannot solve more general problems of distributive justice in the same way we solve the age-group problem. In fact, our prior solutions to other problems of distributive justice must limit or *frame* the age-group problem, much in the way my neighbors' legitimate boundaries frame my property and restrict my landscaping plans. Second, we must determine what *form* or type of prudential reasoning is appropriate to the restricted problem that remains. Specifically, can we use the standard model of a fully informed rational agent who seeks to maximize his expected well-being over his lifetime? Or must we modify that standard model in an important way?

In this and the next two sections, I will address these issues of scope and form. I ask the reader's patience in the remainder of this chapter, for the simple intuition about prudence and the lifespan must be qualified in various ways before it can really be used to solve our

problem. These qualifications will involve some fairly abstract considerations. By the end of the chapter, however, the forest will again
emerge from the trees.

We must restrict or *frame* the age-group problem by drawing on our
prior answers to more basic questions about distributive justice. We
must first know, for example, what principles of distributive justice
govern income or health-care distribution in general before we can
determine how income or health care should be distributed between age
groups. To see why this is true, consider what happens if we try to
appeal to a standard form of prudential reasoning to solve an *unframed*
or unrestricted problem of justice between age groups. For example,
suppose we would like to know how to distribute health-care resources
between the old and the young, but that we do not know what justice
requires in the distribution of health care between persons. That is,
suppose we appeal to prudential reasoning to solve both problems of
distributive justice simultaneously. We shall see, this would be like
trying to budget health care over the lifespan without knowing how
much we have to budget.

One strategy for solving both problems at once appeals to prudential
reasoning by fully informed economic agents.[2] The proposal is that we
rely on market mechanisms to allow people every chance to express
their own prudent preferences about what health care they want at each
stage of their lives. The social task is to make sure that such markets
function properly. Specifically, since we are concerned with health
care, we might look to a market for health-care insurance that offered a
variety of benefit packages. Benefits might be more or less comprehensive, and people could choose a level of health-care protection that
matched their own preferences—balancing their aversion to risk against
the value to them of maintaining health. The packages might be more or
less focused on acute care or on preventive care, as tastes dictated.
Similarly, packages would vary in the rate at which they "saved"
health-care resources for later stages of life, or they might vary in which
kinds of services were made available at different stages of life. Premiums would reflect the relative costs of these packages. People would
buy the packages they deemed prudent, given many facts about their
situations (their values, preferences, resources, family situation, etc.).
This approach, with a vengeance, converts the problem of age-group

2. Discussion in the next few paragraphs is adapted from Daniels (1985, Chapter 5).

distribution of health care into an individual savings problem. A social problem arises only if the markets within which prudent choices are made are defective in some way.

The strengths and weaknesses of this unrestricted or unframed appeal to prudential reasoning are revealed if we consider the way in which a rational consumer might think about the problem of chronic illness or disability. The long-term care such conditions require is a focal point of criticism of the treatment of the elderly. In the United States, chronically disabled or enfeebled persons tend to be institutionalized more frequently and earlier than comparable persons in other systems, for example, in Great Britain or Sweden. Moreover, they are often institutionalized at inappropriate levels of care, and possibly at a higher cost than alternative forms of treatment or service would involve. The incentives for such institutionalization are built into Medicare and Medicaid reimbursements.

The effects of such "overmedicalization" on both the mental and physical health of the elderly are serious (Morris and Youket 1981). Yet, as Christine Bishop (1981) points out, the uncertainty facing the onset and costs of disability make it an obvious candidate for insurance. The rational consumer would presumably try to buy a package that avoided the objectionable features of our current long-term-care system.

Individuals face a significant, actuarially calculable chance of chronic illness or disability over their lifetime; the chance generally increases with age (of course some are disabled from birth or face a known genetic disposition to disability). Although only one in twenty persons over 65 is in a nursing home in a given year in the United States, one in four will at some time enter one (Palmore 1976). Chronic illness or disability may require large expenditures of medical, personal care, or social support services. Moreover, the size of the expenditures for a given disability will vary with other contingencies, such as family situation and preferences for living conditions.

The uncertainty surrounding each of these contingencies and their joint risk suggests that rational consumers will enhance their well-being over their lifetime if they pay a modest insurance premium rather than keep the money and then risk a large loss if they turn out to need such care. Specifically, we might expect rational consumers to want insurance that offered them benefits flexible enough to meet their real needs. They would want alternatives to nursing-home institutionalization if they needed lower levels of care, or some family help, or modifiable living quarters. Thus they would buy contingency claims on the joint risks of disability and other factors, such as the absence of family support or the unsuitability of living arrangements.

The connection between disability as an insurance problem and as a problem of "savings" becomes clear when we see, as Bishop notes, that short-term coverage faces special problems. If coverage is actuarially fair and we pool risks by age, high premiums will face the elderly, those most in need of insurance and those least likely to be able to pay for it because of declining incomes. The prudent consumer, anticipating such higher premiums, would have to save, perhaps by buying an annuity to cover his later premiums. But since no one knows how long he will live, it is hard to predict how much to save. Notice, however, that plans offering lifetime coverage with a fixed premium are equivalent to savings: A community-rated lifetime plan has a built-in savings feature because of the distribution of needs by age.

Although these considerations suggest that there should be a demand for such insurance, we find no actual market offering it. Bishop points to several reasons such insurance is not available: (1) uncertainty about inflation adds to the insurer's risk, where real benefits and not fixed money amounts are involved, so private coverage would be discouraged; (2) administration costs are high and coverage of the population is not extensive; (3) some current public programs would partially undercut the market for such insurance; (4) these forms of insurance are especially vulnerable to "adverse selection," which means too many high-risk people buy, driving premiums up and low-risk people out, and to "moral hazard," which might take the form of people overstating their disability. From these facts, Bishop concludes that private marketing of such insurance is not likely to develop and that we should institute some form of universal, compulsory insurance encompassing medical, personal care, and social support services.

This proposal for compulsory insurance for long-term care is an attempt to derive appropriate social policy from a consideration of what prudent consumers would want to buy, but it also shows why prudential reasoning by fully informed consumers cannot solve the *unframed* or unrestricted age-group problem. A compulsory scheme involves significant income redistribution. Entitlements to benefits, presumably at an adequate or "decent minimum" level, will be subsidized for those who cannot buy them. But why should rational consumers who know they are financially well-off be willing to pay higher premiums or taxes to subsidize those who are poor? If such persons choose to subsidize health care for the poor, their choice will not be dicated by prudence alone. Similarly, fully informed rational consumers who know they are at low risk for certain medical problems would be prudent not to enter insurance schemes that include those at high risk for the same conditions. For example, if I know I have several children whom I am likely to prepare for lucrative careers, I might not want to be in a risk pool with childless people, who are more likely than me to need home care when they are frail and elderly. If I am committed to an insurance scheme that is not actuarially fair to me, because it cross-subsidizes those with higher risks, then I am not motivated by considerations of prudence alone.

We are asking the impossible if we expect fully informed consumers to choose a just distribution of health care (or income) on nothing more than prudential grounds. Problems of distribution are irreducibly interpersonal, wherever we redistribute income across risk groups. In general, when we share—that is, insure—risks across groups that we know have different incidences of disease (like the rich and the poor, blacks and whites, men and women), we redistribute income between persons. In such cases, prudent consumers who know they are in a low-risk group would not be inclined to transfer income to high-risk people. Such choices may leave fairness or justice aside. Therefore, even if prudential reasoning solves the problem of justice between age groups, it cannot solve unframed problems of distributive justice.

One issue about framing the age-group problem must be made explicit. It is central to the claim that some form of prudential reasoning offers us a solution. We must assume that the institutions we are to reason about prudentially operate over our lifespan. If we knew we would live under these institutions only through the late stages of our lives, because we were already old, then it would be prudent for us to authorize substantial transfers from the young to the old. Unless we assume that institutions operate over our lifespan, we introduce an irreducibly *interpersonal* problem of distribution. The problem again becomes one of *them* and *us*. It becomes a problem for which more general principles of distributive justice are needed. The age-group problem is equivalent to a problem of budgeting within a life only if we assume institutions operate over the whole lifespan. The prudent consumers designing it must plan accordingly, that is, without reference to—or knowledge of—their actual age. Only by denying individuals information about how old they are can we retain the frame within which the appeal to prudence may readily be justified.

This motive for restricting knowledge about age, or denying the use of such knowledge, must not be confused with the justification some contractarian philosophers give for restricting knowledge within their constructions. Some of the most important recent work in the general theory of justice involves efforts to use a form of prudential reasoning to solve the purely interpersonal problems of distributive justice which lie outside the frame of our age-group problem (Harsanyi 1976; Rawls 1971). On these approaches, rational agents are set the task of choosing principles to govern social institutions, but they are blinded to all facts that distinguish them as particular individuals.[3] They are able to reason about their well-being only in a schematic way, free from knowledge of their actual preferences. It is a difficult task to show why this hypothetical contract by ignorant agents is the appropriate perspective from which to approach the general or interpersonal problem of justice. For example, Rawls (1980) argues that only appropriately veiled agents can serve as models for the ideal of persons which underlies liberal political theory; the contract must be drawn in a way that is procedurally fair to such agents (cf. Daniels 1979, 1980). These arguments take us well beyond the scope of this book, where we are tackling only the more modest age-group problem.

We have put interpersonal issues of justice outside the frame of our problem by assuming we already have solutions to them. Once restricted in this way, the age-group problem is quite naturally addressed by some form of prudential reasoning. Nevertheless, even this restricted appeal to prudence requires special restrictions on what prudent deliberators may know. I already explained that we must blind rational agents to their age if we are to keep their deliberations within the appropriate frame. But other restrictions on knowledge are also necessary, as we shall see in what follows. I shall argue that it is possible to motivate these restrictions on knowledge without invoking all the considerations needed to motivate the contractarian approach to the full-blown problem of distributive justice. To make good on this claim, I must return to the second issue I raised earlier that needs clarification, that is, the form of prudential reasoning appropriate to the age-group problem.

Prudence: Mine, Yours, Ours

We may appeal to prudence to solve the age-group problem only if we *frame* that problem. We must constrain prudential reasoning about the

_____

3. Gauthier (1986) tries to avoid this device.

age-group problem by assuming that other principles of distributive justice already govern interpersonal distributions. These principles of justice define the overall budget that prudent deliberators must allocate over the lifespan. We must now consider what form of prudential reasoning is appropriate within this frame.

Consider again the problem of the just distribution of health care between the young and the old. Suppose I know I have available to me a lifetime health-care allocation in the form of an insurance benefit package and that this package is my *fair share* of health-care benefits. It is all I am entitled to on grounds of justice. It is a level of health protection delivered by a just health-care system, the age-group problem aside. (For now I will leave unstated what principle of justice might govern the design of health-care institutions. Though I leave indeterminate the notion of a fair share here, I shall return to this issue in the next chapter.) Again let us suppose that prudential reasoning about this insurance package—budgeting it over a lifespan—is undertaken from the perspective of a fully informed rational agent, in this case me. My task is to budget this benefit package, once and for all, so that it is used to meet my needs and preferences over my lifetime. How would it be rational for me to budget it—given all the uncertainties about my future health, wealth, and family situation?

I might reason as follows. I know that I must consider my needs at each stage of my life, for I must live with the choices I make. If I save health-care resources in such a way that I am entitled to whatever it takes to prolong my life when I am old and terminally ill, then I will have to avoid spending these resources when I am young and when having them available could do more for me than merely briefly postponing my death. Therefore it seems prudent for me to reserve certain life-extending technologies for my younger years. I would thus maximize the chances of my living a normal lifespan. I might also use some of the resources "saved" in this way to provide myself with more social-support and home-care services if I turn out to need them in my old age. I might reason that such services could vastly improve the quality of my life in old age and that such an improvement is worth the increased risk of a slightly shortened old age. I would then—through my benefit package—instruct the providers to treat me accordingly, that is to appeal to an age criterion in their utilization decisions concerning me.

Such an insurance package for an individual is intended to resemble some features of the British National Health Service and to contrast with corresponding features of our own system. In the British National Health Service, for example, hemodialysis is not used on elderly patients, who would very likely receive it in the United States. British physicians tend to view elderly patients as "medically unsuitable" for dialysis (though experience in the United States would tend to counter this claim). These expensive resources are reserved for younger patients (Aaron and Schwartz 1984). At the same time, far more extensive home-care services, alternatives to early—and premature—institutionalization of the frail elderly, are available. Since the British have often been criticized for "age bias" in their "unequal" treatment of the elderly, the insurance package I just described, and the rationale I provided for it, are of some interest. If prudence governs the design of individual insurance packages, and if we can generalize from such individual plans to social insurance schemes, then we may be justifying a form of rationing by age that has received sharp moral criticism. I shall return to the issue of rationing by age in Chapter 5. A more pressing issue about the appeal to prudence must be faced first.

The simple argument I have just sketched, generalizing from an insurance package that is prudent for me to the design of a package or institution appropriate for society, is open to an objection like the one that occupied us earlier in the chapter. There is no one prudent plan for all individuals because what is prudent is determined by an individual's preferences, resources, and individual needs. If Norman Daniels gives explicit consent, after prudent deliberation, to being treated differently at different points in his life, even denying himself dialysis when he is old, then that is one, relatively uncontroversial, thing. But other individuals may have different preferences. Prudence might very well dictate other insurance schemes for them. They may reject differential treatment by age or they may prefer other uses of age criteria. Ordinarily, prudent choices will reflect full knowlege of a person's situation and conception of what is good in life. There is no one insurance package all would agree is "the prudent one." There are many prudent ways to budget our fair shares of health care.

One response to this objection amounts to an attempt to let us have our cake and eat it too. Suppose people could choose among several lifetime health-insurance packages, each of which represented a different conception of prudent lifetime allocation. We might even imagine that our publicly subsidized health-insurance programs, such as Medicaid and Medicare, had alternative benefit plans and people could choose which would apply to them. Then no individual's judgments of prudence would be imposed on anyone else, yet prudence would play a role in the design of the health-care system.

This proposal sounds fine until we realize that we would have to bind people to the choices they made at some very early point in their adult lives. If we allowed people to change insurance plans as they aged, we would create conditions that would bankrupt the plans (cf. Bishop 1981). We could not, for example, let young people buy into or join plans that steered resources toward their young years at the expense of their later years, and then jump, when they were older, into plans that skewed allocations in the opposite direction. Letting people get the best of both arrangements would dramatically increase the benefit package—and its costs—overall. Yet, we were supposing that the overall package was an individual's fair share of health care within a just health-care system. Plan-jumpers would be exceeding that fair share.

Requiring a once-and-forever choice by young adults, in order to protect the system against bankruptcy, would *bias* the plans toward the interests and preferences of young adults. As an individual ages, his situation changes. The passage of time erases uncertainties and alters probabilities that must affect early judgments about what it is prudent to do, even assuming preferences and values remain constant. But these are likely to change too.

As individuals age, they may revise their conceptions of what is good in life. They may revise their plans of life in ways that reflect both what they learn and what they come to adopt as fundamental goals and values. What is prudent from the perspective of an early plan of life may not be from the perspective of a later one. To a young person anticipating raising a large family, it might seem prudent to choose a plan which deemphasized long-term care late in life in favor of one that provided generous acute care. But some time later, disappointment in marriage or career could change expectations about family size and thus about the prudent design of the insurance package. Letting each person choose a plan early in life fails because it gives undue weight to the choices made at a particular, early point in life. We avoid imposing the prudence of one individual on another only by imposing the prudence of the young on the old.

Thus we can see that adhering to the prudent choices of fully informed consumers risks biasing our health-care system in favor of the prudence of the young. Indeed, practical attempts to increase the role of individual prudential reasoning in medical treatment and resource-allocation decisions rely on strategies such as "precommitment" or "advance directives" (Living Wills) or the purchase of long-term insurance plans. All of these encounter the problem of potential age-bias. I shall return to the practical manifestations of this problem later when I discuss rationing acute care and planning for long-term care. From a theoretical perspective, however, there may be a way to avoid the issue of bias toward the young. In what follows I shall propose a strategy for solving the age-group problem that does not rely on fully informed consumers of health-care plans. On this strategy, prudential reasoning will be neutral between the interests of the young and the old.

Veiled Prudence

In the preceding section I pointed out a serious difficulty for our project if we adhere to the standard economist's model of prudential reasoning, which uses fully informed consumers. Once we notice that what is prudent for one fully informed person may not be for another, we try to find a way to avoid imposing the prudence of one individual on others. The only way to do that, however, is to require such consumers to choose early in life how they will budget their fair share of health care. The serious problem with this approach is that it respects the individuality of the young at the expense of the old. The price of respecting the difference between persons is a form of age-bias. To avoid this problem we shall have to retreat from the perspective of the fully informed rational consumer.

This retreat from full information can be motivated to a significant extent by considering the demands of prudence itself. Traditionally, the requirement that an individual should be concerned with his well-being over his lifetime (cf. Sidgwick 1907) has been a central principle of the theory of individual rationality. That is, we must each assess our well-being in a way that is *neutral* with regard to time. What counts is the quality of our experience, not when in our lives it takes place. We must not, for example, "discount" the value of our experience or well-being merely because it occurs later in life rather than earlier.

Recently, this classical principle has been challenged (Parfit 1984), and it is now more controversial than it has traditionally been. In the

Appendix to this volume I discuss this important controversy in some detail. I conclude that the strategy I pursue in this chapter, including the standard assumptions I make about prudence, is not rendered unacceptable by these recent attacks on the classical account. The philosophically inclined reader will eventually want to assure himself that this controversy is only a detour from my main project.

Since I am persuaded this controversy is but a (scenic) detour, I shall assume in what follows that a *time-neutral* concern for well-being over the lifespan is one of the demands of prudence itself. I shall also assume that we will not discount well-being in proportion to the degree of psychological connectedness that holds among the different stages of our lives (cf. Parfit 1984). Prudence will not allow us to say, "In the future, I will no doubt be quite different from what I am like now, and so I now care much less about what will happen to me then than I care about what happens to me in the near future." In political philosophy, including the theory of justice, since we are concerned with the design of institutions operating over the lifespan, the classical assumption that we are equally concerned about all parts of our lives is usually made. Moreover, this assumption begs no *practical* questions that people currently pose. For example, baby boom opponents of our Social Security system who want to "privatize" it, eliminating transfers from the current young to the current old, are not acting on the assumption that they care little about their future well-being. Rather, their concern is to protect their future from a transfer they believe is unfair to their birth cohort.

How does a concern for my lifetime well-being motivate restrictions on full information? Suppose I have full information about myself and my current plan of life and that I want to plan the allocation of basic resources over my lifetime. I am aware that my own preferences and values have changed as I have matured and aged. What was important to me years ago is in some cases not important now. What I could not then imagine to be desirable is now important to me. As I think about my future, I should learn from the past. My conception of what is good in life changes, though not always suddenly or dramatically. Over time, these changes can be fundamental. If I am concerned about my well-being in the future, I must be concerned about what I will then think is good, and not just what I now think is. Yet, I cannot know in detail what my future conception of what is good will be. Nor can I extrapolate merely from what I value now or from how I have changed in the past. The dilemma I face is that I must respect both my current plan of life and the ones I may come to have, even if I am not sure of their details.

The key to reasoning prudently about this problem is the idea of keeping my options open. Though I want to pursue my current projects and what I now think is good to do and be, I cannot prudently ignore the fact that I may think differently about what is good in the future. To keep options open, I must ensure that basic resources and opportunities are available throughout my life. I do not want to deny my future self the chance to pursue what I shall come to think as valuable and important. In keeping options open, however, I cannot merely think about them from the perspective of my current plan of life. Prudence itself demands that I step outside of my current conception of what is good in life and keep options open that are not of primary importance to me from my current perspective.

Because this demand to keep options open derives from my prudent concern for my well-being over my lifetime, I need a perspective from which I can deliberate about lifetime well-being. I must be able to abstract from or disregard the particulars of my conception of what is good at any given time, including the present, when I consider my whole life neutrally. That perspective cannot just be my current view of what is good for me.[4] "Keeping options open" means that I want to assure myself at each stage of life of having an adequate chance to pursue whatever my plan of life is then. I must assure myself that at each stage of life I shall have a reasonable share of basic social goods which serve as the all-purpose means of pursuing what I think is good. Keeping options open implies I must be neutral or unbiased toward the different stages of life that I shall go through.

The idea that prudence is concerned with lifetime well-being thus drives us beyond the perspective of the fully informed rational consumer or choice-maker who deliberates from within a given plan of life. Still, there is an important tension between prudence, seen in this way, and our commitments to our goals which are defined within our current plan of life. Prudence tells us to keep options open and not to throw

4. Brandt (1979) argues that we cannot have a coherent perspective from which to think about the satisfaction of desires over a lifetime. He uses this point to argue that we should adopt an "enjoyment" theory of happiness, rather than a desire-satisfaction theory. My view retreats—for purposes of lifetime prudential planning—to the perspective of primary social goods.

caution to the wind in pursuit of what we currently think it is good to do or be. Many of the best things in life, however, come about only if we are fully committed to the important projects within our current plans of life. Moreover, we are trying to use prudential thinking about resource allocation over a lifetime as a model for thinking about the just distribution of resources between the young and the old. If prudence threatens commitment, then it may also threaten individual autonomy when we translate its lessons into talk about justice. These are important issues best discussed after we have a clearer idea how the Prudential Lifespan Account is to be fleshed out.

There is an important objection to my claim about what prudence demands. I have said that a concern for our well-being over our lifetime implies that we must keep options open. We must make sure that basic goods at each stage of life are as nearly adequate as possible to pursuing the plan of life we will have then. The objection is that this strategy may seem unduly risk-averse; it may only express a particular aversion to risk. It is what prudence demands only for those unwilling to take the risk of living their current plan of life with true commitment.

Pursuing a particular plan of life without keeping options open, however, is more than a willingness to take risk. It may be a form of dogmatism about what is good.[5] Conceptions of what is good are *incommensurable* in the sense that we have no perspective from which we can neutrally rank them. We have no view of the good life that transcends all others and by reference to which we can measure or rank all others. Rawls (1971, 1982) has made this point central to his discussion of why, for matters of justice, we must avoid appealing to the notion of utility or satisfaction as a basis for ranking conceptions of what is good. It is striking that a similar point can be made from the perspective within a life over time and not only about comparisons about the plans of life of different persons. What prudence is telling me is that I must be tolerant of the plans of life I may come to have, and indeed have had in the past! I cannot in general rank two conceptions of what is good that I hold at different times because I have no perspective from which to make such rankings.

This point may be reflected in a judgment we sometimes make. When we notice that something that was once very important to us no longer is, we may be struck by the difference in ourselves. We may even say,

5. See Buchanan (1975) for a similar point.

"I was different then, not necessarily better or worse, just different."[6] We might not even be able to say that there is a difference in happiness or satisfaction, at least no difference that makes us say, "I'd rather be the way I was then" or "I prefer the way I am now." I am not suggesting that changes in plans of life are without reasons, even good reasons. Still, even if we have good reasons for revising our plans of life, it does not mean that we can judge one plan to be better than another in some global sense. Insisting that we treat our current plan of life as the only perspective from which to judge lifetime well-being thus appears to be a kind of intolerance and dogmatism.

There is another way to pursue the objection that keeping options open is too risk-averse a strategy: We might reject the maxim that we "keep options open" because we think significant changes in life plans are *rare*. Accordingly, it would be wrong to treat such an atypical feature of life as a reason for keeping options open. Rather, since the likelihood of change is so low, it is reasonable to view lifetime well-being only through the lens of our current plan of life. If we follow the odds, which favor the persistence of our current perspectives, we have the best likelihood of promoting lifetime well-being. The person who insists on keeping options open thus may appear to be obsessively focused on avoiding outcomes that have extremely low risk.

Two points can be made in response. First, this version of the objection rests on a mistaken view about stasis in our lives. To say that our views about what we think it is good to do or be are static ignores the fact that we are trying to allocate resources over a full lifespan. Given that time frame, the probability that people are frozen into the perspectives they have in youth is low. We are being asked to ignore much of what we know about maturation and aging. Second, many contingencies we face over a lifespan involve true uncertainties and not mere probabilities. Our well-being at any time depends crucially on our ability to adjust our plans of life to the uncertainties we encounter. This adjustment is not just passive reaction, but a way of *controlling* our lives. Looking at our lifetime well-being only from the perspective of our current plan of life ties our hands in the face of real uncertainties. It reduces our capacity to make the new decisions living requires. In sum,

6. We might even *say*, "I was a different person then," meaning that we then thought quite differently about things. But plans of life may change without implying the discontinuity of persons. See the Appendix.

rejecting the strategy of keeping options open either is a form of dogmatism, or rests on an erroneous estimate of stasis in our lives, or is based on a confusion about how to quantify important contingencies (as risks or true uncertainties) over the lifespan.

To specify further the form of prudential reasoning needed to solve the age-group problem, I shall borrow from the contractarian approach developed by Rawls (1971, 1980) for the general problem of justice. In thinking about my well-being over my lifetime, prudence demands that I abstract from the perspective of my current plan of life, including my knowledge of my stage of life. One way to do this is to imagine that I know I shall have a conception of what is good in life, but that I do not know exactly which one it is. I may think of myself as *free* to form and revise my plan of life over time, and my lifetime well-being will depend on my having available to me at least a fair share of the all-purpose means for pursuing my ends, whatever they turn out to be.[7] In Rawls's work, contractors are to measure their well-being by using an index of primary social goods. They are to think of their well-being as being determined by their *lifetime expectation* of having a certain distribution of basic liberties, opportunity, income and wealth, and the social basis for self-respect. Their task is to pick principles of distributive justice that govern the design of basic institutions. Specifically, they are to think about which inequalities in well-being (as measured by this index) they might accept as rational, under the various motivational and information constraints imposed by the circumstances in which they are to make their choice.

In our restricted or framed problem, the problem of justice between age groups, we are to suppose that hypothetical deliberators already know what inequalities in lifetime expectations are acceptable. That is, they already know what counts as a fair share, and they must decide a

---

7. Gibbard (1982) defends a method of reasoning prudentially about features of our health-care system in accordance with an ex ante pareto principle. He notes, however, that the perspective from which such reasoning should take place is ex ante our knowlege of many of our particular individual traits, indeed, from a perspective prior to conception. What is not at all clear is how such a prudential judgment works: What notion of well-being is being maximized over the lifetime? Dworkin (1986) also advocates a prospectively planned insurance approach which operates over the lifespan. Both Gibbard and Dworkin appear more willing than I am to let prudence answer more general questions about distributive justice, for example, for health care in general, and not just for the age-group problem.

further question: How *should* that lifetime expectation of enjoying a certain level of primary social goods be distributed over each stage of life so that lifetime well-being is maximized? Even if we have solved the general problem of distributive justice, we have not solved the age-group problem until we answer this question.

It is important to see that I can here borrow some of Rawls's apparatus for solving the general problem of justice without having to import his whole justification for its features. His argument that reasonable people would accept the restrictions of the contract position, as a device that is fair to all parties, depends on acceptance of certain Kantian views or ideals about the nature of moral agents. Specifically, the setting for a contract that Rawls develops is intended to capture the idea that moral agents have two fundamental moral powers, making them equal and free. They are *equal* in the sense that they are motivated to abide by considerations of justice, and they are *free* in their capacity to form and revise conceptions of what is good. Rawls believes that constraints on choice, such as the "thick veil of ignorance" he imposes and the "thin theory of the good," that is, his index of primary social goods, are fair to persons because they reflect this basic ideal about the nature of moral agents. It is reasonable for us to abide by the hypothetical contract because, presumably, we share these background ideals about persons and the role of morality in society. At least within the liberal democratic tradition, these are a shared ideal (Rawls 1980).

My use of Rawlsian devices does not depend on an appeal to his robust Kantian account of the nature of persons or to his claims that the choice situation is procedurally fair to such persons. Thus my justification rests on more modest grounds—herein lies both its strength and the restrictions on it. I have argued that prudence itself (under the standard assumptions) requires that individuals respect their own changes in their conception of what is good at each stage of life. Their concern for their own lifetime well-being will require them to abstract from full information in order to be neutral about each stage of their lives, at least when they are considering the design of institutions that affect them over the whole lifespan. The price to be paid for resting my justification on this more modest ground is that we can use prudential reasoning to solve only a more modest problem. My prudent deliberators, even though they use Rawlsian restrictions on prudential reasoning, cannot attempt to solve problems of justice which cross the boundaries between persons. My prudent deliberators are concerned only with the *framed* problem of justice between age groups.

Summary

The argument of this chapter has been complex and a summary of what I have shown will help us move on. If we treat people differently by race or sex, then we risk violating principles governing equality among persons. Treating the young and old differently, however, may not mean treating people unequally. Over a lifetime, such differential treatment may still result in our treating people equally. From the perspective of institutions that operate over our lifespan, transfers *between* age groups are really transfers *within* lives. Transfers that look like subsidies by the young of the old are transformed into a type of savings. Indeed, prudent allocations of resources over a lifetime will have the desirable characteristic that they benefit everyone passing through these institutions as they age. If we can determine what form of prudential reasoning we should use to design such institutions, then we may specify which transfers between age groups actually improve prospects for everyone. We will be using what is prudent over a lifespan to determine a result that neither young nor old can object to as unfair to them.

We cannot, however, rely on prudential reasoning by fully informed rational agents, the standard model of the welfare economist, to carry out our task. First, we must restrict the scope of the appeal to prudential reasoning. We must assume that we are appealing to prudence only within the frame provided by a set of principles that govern distributive justice between persons. Prudential reasoning by fully informed persons cannot tell us about what is just when we transfer goods in a way that irreducibly crosses the boundaries between persons. Put more simply, for prudent budgeting, a budget is needed, but what the budget is must depend on what is just, on what people are entitled to. An important aspect of framing our problem is to insist that we are concerned with institutions that operate over the whole lifetime; hence we must blind our prudent planners to their age. We must also modify the form of prudential reasoning for other reasons. If we set fully informed consumers the task of budgeting their fair share of health care over their lifetime, we would have to restrict them to choices made at an early point in life. Otherwise, jumping from plan to plan would lead them to exceeding their fair share. But then we seem to bias plans in favor of what the young take to be prudent and ignore the prudence of the old.

More careful thought about what lifetime prudential planning requires leads us to a solution to this problem of bias. If we are concerned about well-being over a lifetime, we must make sure that at each stage of our lives we have available to us resources which can serve as the all-purpose means for pursuing whatever our plan of life at that stage happens to be. The device of putting our consumer behind a Rawlsian veil of ignorance that blinds him to any knowledge of his particular conception of what is good, and asking him to think instead about well-being in terms of an index of primary social goods, gives us a way of making the reasoning about lifetime well-being appropriately neutral. The reasons for appealing to this veiled form of prudence derive, however, from the requirements of prudence alone.[8] Arguments about the ideal nature of persons, like those in Rawls's justification of his contractarian approach to the general problem of distributive justice, are not needed here, as we are concerned with a more narrow problem.

Simply stated, then, we can solve the age-group problem by considering what principles for the design of institutions prudent planners would choose to govern the institutions they pass through as they age. These planners must make their choices, however, from behind a veil of ignorance that keeps them from knowing their age or their conception of

8. This point is overstated as it stands, as R. M. Hare has pointed out to me. No account of prudence should imply that agents should *always* ignore certain or fixed facts about themselves, such as that they are females or hemophiliacs, when they make prudent choices. Ignoring such facts will lead to erroneous choices in many contexts. Invoking a veil of ignorance, however, as I have done above, seems to mean blocking out knowledge of such facts. So prudence could not plausibly require that in general we make choices from behind such a veil. Nor have I argued that it does. Rather, I have argued that prudence gives us general reasons to invoke a veil as a way of avoiding age-biasing our choices in certain special contexts, for example, in designing institutions which distribute important goods to us over the lifespan. In my construction, we already depart from purely prudential considerations, that is, from the standard model of the fully informed agent, when we block out knowledge of age. Remember that our deliberators must assume they will live through the whole lifespan. This part of the veil of ignorance is needed to keep the "frame," which excludes interpersonal issues, intact.

what is good in life. They can judge their well-being only by reference to an index of primary social goods. Moreover, the principles they choose must fall within the frame imposed by more general principles of distributive justice. This is a very abstract formulation of the method for solving the age-group problem; I shall provide a more concrete discussion in the next chapter.

4

Justice in Health Care

The Prudential Lifespan Account

To solve the problem of justice between age groups we must solve a substitute problem: How would rational agents design institutions to prudently allocate fair shares of basic social goods over their lifespan? In substituting a problem of individual rational choice for a problem of social choice, our strategy resembles recent contractarian approaches to the general theory of justice. Rawls (1971), for example, sets contractors the task of solving a specially constructed individual rational choice problem in his Original Position. The solution to this hypothetical choice problem is supposed to yield principles of social justice. This strategy raises a basic question about justification: Why is it reasonable to accept the individual choice problem as a method for solving the social problem of selecting principles of justice? Rawls's answer is that his construction adequately reflects ideals which we share, at least within the liberal democratic tradition, about the nature of persons and of the role of morality in society (cf. Rawls 1980, 1985; Daniels 1980). Much of the controversy surrounding his theory concerns this justification for substituting one problem for another.

In the case of the age-group problem, our substitution of a problem of prudence for one of social choice requires no such elaborate and controversial justification. Because of the sobering fact that we age, there is a natural convergence of the age-group problem, appropriately "framed," and our substitute problem of prudent allocation over a lifespan. From the perspective of institutions that distribute basic goods over our lifespan, transfers between age groups are equivalent to transfers within a life.[1] The substitution of one problem for another is appropriate because, in the peculiar case of age groups, they are essentially the same problem. Of course, we must modify the standard model of prudential reasoning by fully informed persons. We must impose a veil of ignorance which blots out knowledge about one's age and about one's plan of life. This superficial resemblance to Rawls's construction should not confuse us, however, since the justification for these restrictions is different from his. The restriction on knowledge of one's age is a way of keeping our prudent choice problem within the frame that excludes transfers across the boundaries between persons. The other restrictions on knowledge follow from considerations of prudence over the lifespan, not from an attempt to model an underlying ideal of the nature of moral agents.

To solve the age-group problem, then, we must solve a problem of prudent choice that has several specific features. First, our rational deliberators must seek principles to govern the design of institutions that distribute basic resources over the lifespan. Second, these deliberators must already know that the basic goods being distributed constitute a fair or just share, that is, that more general principles of distributive justice already solve problems of distribution between persons. Third, the deliberators must assume that they will live through each stage of life under the institutions they are designing, thus avoiding conversion of the problem into one that involves transfers between persons or between birth cohorts. Fourth, the deliberators must also not know their plan of life or conception of what is good in life; instead they must measure their well-being by reference to a Rawlsian index of primary social goods. The justification for this final restriction derives from requirements of prudence, for it would not be prudent to bias the design of institutions in favor of choices thought prudent only by the young, for example (see Chapter 3, p. 57ff., and p. 64, n.8).

This characterization of the Prudential Lifespan Account is very abstract, and it is my task in this chapter to make it more concrete.

---

1. This claim involves an idealization in which the age-group and birth-cohort questions are kept sharply distinct. This idealization is examined more critically in Chapter 7.

Specifically, I shall explain in some detail how the account would be filled out in the case of health care. I shall save my discussion of income support for Chapter 7. My first task will be to give some explanation of what distributive justice in general requires in the case of health care, an account which I have developed in much greater detail elsewhere (Daniels 1985).

Equal Opportunity and Health Care

Prudent deliberators solving the age-group problem work within a frame. They must choose principles governing the design of institutions that distribute *fair shares* of basic social goods over the lifespan. To see how their deliberations about health care will go, however, we must fix the notion of a fair share of health-care resources. Specifically, we want to know what principle of justice determines such shares. For purposes of illustration, I could have chosen any principle of justice for health care and shown how prudent deliberators could address their allocation problem using it as their frame. In subsequent chapters, however, I shall want to draw substantive moral conclusions in which I have some confidence. Therefore, I shall frame the age-group problem for health care with an account of just health care I have developed elsewhere.[2]

A natural place to seek principles of justice for regulating health-care institutions is by examining different general theories of justice. Libertarian, utilitarian, and contractarian theories, for example, each support more general principles governing the distribution of rights, opportunities, and wealth, and these general principles may bear on the specific issue of health care. But there is a difficulty with this strategy. In order to apply such general theories to health care, we need to know what kind of a social good health care is. An analysis of this problem is not provided by general theories of justice. One way to see the problem is to ask whether health-care services, say personal medical services, should be viewed as we view other commodities in our society. Should we allow inequalities in the access to health-care services to vary with whatever economic inequalities are permissible according to more general principles of distributive justice? Or is health care "special" and

2. I draw on Daniels (1985, Chapters 2–3) in the remainder of this section.

not to be assimilated with other commodities, like cars or personal computers, whose distribution we allow to be governed by market exchanges among economic unequals?

Is health care special? To answer this question, we must see that not all preferences individuals have—and express, for example, in the marketplace—are of equal moral importance. When we judge the importance to society of meeting someone's preferences we use a restricted measure of well-being. We do not simply ask, how much does the person want something? Or, how happy an individual will be if he gets it? Rather, we are concerned whether the preference is for something that affects well-being in certain fundamental or important ways (cf. Scanlon 1975). Among the kinds of preferences to which we give special weight are those that meet certain important categories of need. Among these important needs are those necessary for maintaining normal functioning for individuals, viewed as members of a natural species. Health-care needs fit this characterization of important needs because they are things we need to prevent or cure diseases and disabilities, which are deviations from species-typical functional organization ("normal functioning" for short).

This preference suggests health care may be special in this restricted sense: Health care needs are important to meet because they affect normal functioning. But there is still a gap in our answer: Why give such moral importance to health-care needs merely because they are necessary to preserve normal functioning? Why is preserving normal functioning of special moral importance? The answer lies in the relationship between normal functioning and opportunity, but to make the relationship clear, I must introduce the notion of a normal opportunity range.

The *normal opportunity range* for a given society is the array of life plans reasonable persons in it are likely to construct for themselves. The normal range is thus dependent on key features of the society—its stage of historical development, its level of material wealth and technological development, and even important cultural facts about it. This dependency is one way in which the notion of normal opportunity range is socially relative. Facts about social organization, including the conception of justice regulating its basic institutions, will also determine how that total normal range is distributed in the population. Nevertheless, that issue of distribution aside, normal functioning provides us with one clear parameter affecting the share of the normal range open to a given individual. It is this parameter that the distribution of health care affects.

The share of the normal range open to individuals is also determined in a fundamental way by their talents and skills. Fair equality of opportunity does not require opportunity to be equal for all persons. It requires only that it be equal for persons with similar skills and talents. Thus individual shares of the normal range will not in general be *equal*, even when they are *fair* to the individual. The general principle of fair equality of opportunity does not imply leveling individual differences. Within the general theory of justice, unequal chances of success which derive from unequal talents may be compensated for in other ways. I can now state a fact at the heart of my approach: Impairment of normal functioning through disease and disability restricts individuals' opportunities relative to that portion of the normal range their skills and talents would have made available to them were they healthy. If individuals' fair shares of the normal range are the arrays of life plans they may reasonably choose, given their talents and skills, then disease and disability shrinks their shares from what is fair.

Of course, we also know that skills and talents can be undeveloped or misdeveloped because of social conditions, for example, family background or racist educational practices. So, if we are interested in having individuals enjoy a fair share of the normal opportunity range, we will want to correct for special disadvantages here too, say through compensatory educational or job-training programs. Still, restoring normal functioning through health care has a particular and *limited* effect on an individuals' shares of the normal range. It lets them enjoy that portion of the range to which a full array of skills and talents would give them access, assuming that these too are not impaired by special social disadvantages. Again, there is no presumption that we should eliminate or level individual differences: These act as a baseline constraint on the degree to which individuals enjoy the normal range. Only where differences in talents and skills are the results of disease and disability, not merely normal variation, is some effort required to correct for the effects of the "natural lottery."

One conclusion we may draw is that impairment of the normal opportunity range is a (fairly crude) measure of the relative importance of health-care needs, at least at the social or macro level. That is, it will be more important to prevent, cure, or compensate for those disease conditions which involve a greater curtailment of an individual's share of the normal opportunity range. More generally, this relationship between health-care needs and opportunity suggests that the principle that should govern the design of health-care institutions is a principle guaranteeing fair equality of opportunity.

The concept of equality of opportunity is given prominence in Rawls's (1971) theory of justice, and it has also been the subject of extensive critical discussion. I cannot here review the main issues (see Daniels 1985, Chapter 3), nor provide a full justification for the principle of fair equality of opportunity. Instead, I shall settle for a weaker, conditional claim, which suffices for my purposes. Health-care institutions should be among those governed by a principle of fair equality of opportunity, provided two conditions obtain: (1) an acceptable general theory of justice includes a principle that requires basic institutions to guarantee fair equality of opportunity, and (2) the fair equality of opportunity principle acts as a constraint on permissible economic inequalities. In what follows, for the sake of simplicity, I shall ignore these provisos. I urge the fair equality of opportunity principle as an appropriate principle to govern macro decisions about the design of our health-care system. The principle defines, from the perspective of justice, what the moral *function* of the health-care system must be—to help guarantee fair equality of opportunity. This relationship between health care and opportunity is the fundamental insight underlying my approach.

My conditional claim does not depend on the acceptability of any particular general theory of justice, such as Rawls's contractarian theory. A utilitarian theory might suffice, for example, if it were part of an ideal moral code, general compliance with which produced at least as much utility as any alternative code (cf. Brandt 1979). That utilitarian theory could then be extended to health care through the analysis provided by my account. Because Rawls's is the main general theory that has incorporated a fair equality of opportunity principle, I have elsewhere suggested in some detail (Daniels 1985, Chapter 3) how it can be extended, with minor modifications, to incorporate my approach. These details need not distract us here.

The fair equality of opportunity account has several important implications for the issue of access to health care. First, the account is compatible with, though it does not imply, a multitiered health-care system. The basic tier would include health-care services that meet health-care needs, or at least important needs, as judged by their impact on opportunity range. Other tiers might involve the use of health-care services to meet less important needs or other preferences, for example, cosmetic surgery. Second, the basic tier, which we might think of as a "decent basic minimum," is characterized in a principled way, by reference to its impact on opportunity. Third, there should be no obstacles—financial, racial, geographical—to access to the basic tier. (The account is silent about what inequalities are permissible for higher tiers within the system.) Social obligations are focused on the basic tier.

The fair equality of opportunity account also has implications for issues of resource allocation. First, I have already noted that we have a crude criterion—impact on normal opportunity range—for distinguishing the importance of different health-care needs and services. Second, preventive measures that make the distribution of risks of disease more equitable must be given prominence in a just health-care system. Third, the importance of personal medical services, despite what we spend them on, must be weighed against other forms of health care, including preventive and public health measures, personal care and other long-term-care services. A just distribution of health-care services involves weighing the impact of all of these on normal opportunity range. This point has specific implications for the importance of long-term care, but also for the introduction of new high-cost technologies, such as artificial hearts, which deliver a benefit to relatively few individuals at very great cost. We must weigh new technologies against alternatives and judge the overall impact of introducing them on fair equality of opportunity— which gives a slightly new sense to the term "opportunity cost."

This account does not give individuals a basic right to have all of their health-care needs met. Rather, there are social obligations to provide individuals only with those services that are part of the design of a system which, on the whole, protects equal opportunity. If social obligations to provide appropriate health care are not met, then individuals are definitely wronged. Injustice is done to them. Thus, even though decisions have to be made about how best to protect opportunity, these obligations nevertheless are not similar to imperfect duties of beneficence. If I could benefit from your charity, but you instead give charity to someone else, I am not wronged and you have fulfilled your duty of beneficence. But if the just design of a health-care system requires providing a service from which I could benefit, then I am wronged if I do not get it.

The case is similar to individuals who have injustice done to them because they are discriminated against in hiring or promotion practices on a job. In both cases, we can translate the specific sort of injustice done, which involves acts or policies that impair or fail to protect opportunity, into a claim about individual rights. The principle of justice guaranteeing fair equality of opportunity shows that individuals have legitimate claims or rights when their opportunity is impaired in particular ways—against a background of institutions and practices which protect equal opportunity. Health-care rights on this view are thus a species of rights to equal opportunity.

The scope and limits of these rights—the entitlements they actually carry with them—will be relative to certain facts about a given system. For example, a health-care system can protect opportunity only within the limits imposed by resource scarcity and technological development for a given society. We cannot make a direct inference from the fact that an individual has a right to health care to the conclusion that this person is entitled to some specific health-care service, even if the service would meet a health-care need. Rather, the individual is entitled to a specific service only if it is or ought to be part of a system that appropriately protects fair equality of opportunity.

Our remaining task is to show how this fair equality of opportunity account of just health care works as a frame for the age-group problem. A mild caution is due, however. This metaphor of the frame could be misleading in two ways. First, in setting up the notion of a frame, I suggested that fair shares of health-care resources are to be allocated prudently over the lifespan. But these shares are not really a fixed quantity of goods and services. They are the entitlements an individual should have, given one's health status and given a health-care system designed so that it protects fair equality of opportunity. At the risk of stretching the metaphor, I would say the frame is not as rigid as talk about "fair shares" might suggest. Second, as we shall see, we must make some adjustments to the frame in fitting it to our problem. That is, we actually modify the equal opportunity account slightly in solving the age-group problem. This modification is really only the finish we must apply to the fair equality of opportunity account if it is to be used in the real world, in which, alas, we age.

Equal Opportunity Throughout Life

To solve the age-group problem for health care, we must work within the frame created by the fair equality of opportunity account. Rational deliberators must consider what principles are needed for the prudent design of institutions that distribute fair shares of health care over the lifespan. Before examining how they would think about the choice they must make, we need to refine the notion of a normal opportunity range.

Life plans generally have stages that reflect important divisions in the life cycle. Without meaning to suggest a particular set of divisions as a rigid framework, it is easy to observe that lives have phases in which different general goals and tasks are central: nurturing and training in childhood and youth, pursuit of career and family in adult years, and the completion of life-long projects in later years. Of course, what is reasonable to include in a life plan for a stage of one's life depends not only on one's own talents and skills, tastes, preferences, and values, but also in part on social policy and other important facts about a society. Still, within a society, we can modify our notion of the normal opportunity range to make it age- or stage-relative. We can disaggregate our original notion and talk about an individual's fair share of the normal opportunity range at each stage of life. As we shall see, it is this notion of an *age-relative normal opportunity range* that will be important to our prudent deliberators as they seek to apply the fair equality of opportunity account over the lifespan.

I should note that the concept of opportunity we are working with here is a *broad* one. It incorporates concerns that go beyond the *narrow* notion of mobility with respect to jobs and offices. In this way it involves a modification of the fair equality of opportunity principle advocated by Rawls (1971). This broader notion has the drawback of being more vague than the narrower one, but it is clearly necessary to broaden the concept in discussing the importance of health care. It is also necessary if we are to leave room for the concept of opportunity to play an important role over the lifespan and not just in certain stages of life.[3]

Consider the problem facing designers of a health-care system that operates over the lifespan. We can now make the constraints on their problem of choice more specific. First, they seek a principled way to design health-care institutions that distribute fair shares of health-care services over everyone's lifespan. Second, they work within the frame specifically determined by the fair equality of opportunity account. The

---

3. See Daniels (1985:50ff) for further discussion of the broad and narrow concepts of opportunity.

lifetime fair shares of health care designers seek to allocate prudently are to be understood as entitlements to those types of health-care services a just system, as defined by that account, would be obliged to provide. Third, the frame also requires that resource transfers are between stages of a life, not between persons. To accommodate this requirement, we make our rational choosers assume they might live through each stage of life under the principles they choose. In effect, they are ignorant of their age: They cannot assume they have already lived through certain stages or that they will die young and not live through later ones. Fourth, we are assuming that they are concerned about their well-being over their whole lifespan. This assumption about the nature of prudence means they must not allow their deliberations to be biased in favor of what seems prudent merely from the perspective of the plan of life they have at a particular point in life, say early adulthood. To protect against this form of bias, we keep the rational agents from knowing their own plan of life, their own conception of what is good in life. Instead, we borrow a device from Rawls (1971), his "thin theory of the good," and require them to measure their well-being by reference to an index of such primary social goods as basic liberties, opportunity, and income.[4]

Although deliberators behind the veil of ignorance do not know their age or plan of life, they do have knowledge about their society and about health-care technology. For example, they must know important facts about the disease/age profile for their society or they cannot begin to make prudent choices about allocation over the lifespan. They must also know about its level of technological development and its pattern of economic growth. They must also know important demographic facts about their society, for example, that expected longevity has been increasing at certain rates and for certain reasons. And they must know about basic economic and sociological trends that have a bearing on these demographic facts. For example, they would have to know about patterns of family support for frail elderly parents, about participation of women in the work force, and about patterns of family stability and mobility.

Under all these constraints, prudent deliberators would reason as follows about how to adapt the fair equality of opportunity account to

4. There are important problems with Rawls's account of these goods, but for his most recent version, see Rawls (1982) and Daniels (1985:42ff).

the age-group problem. From their perspective, prudent deliberators do not know what their individual situation is or what preferences or projects they might have at a given stage of their lives. Still, they do know that they will have a particular plan of life, indeed, possibly different ones at different stages of their lives, and that this plan of life defines what is meaningful for them. This means that it is especially important for them to make sure social arrangements give them a chance to enjoy their fair share of the normal range of opportunities open to them at each stage of life. This protection of the age-relative normal opportunity range is doubly important because they know they may want to revise their life plans. Consequently, they have a fundamental interest in guaranteeing themselves the opportunity to pursue such revisions. But impairments of normal functioning clearly restrict the portion of the normal opportunity range open to individuals at any stage of their lives. Consequently, health-care services should be rationed throughout a life in a way that respects the importance of the age-relative normal opportunity range. In effect, all specific allocation decisions must be constrained by this principle.

I noted earlier in the chapter that the metaphor of a *frame* might be misleading. The reasoning I have just attributed to our prudent deliberators makes explicit how the frame has to be modified to solve the age-group problem. In general, health-care institutions should be governed by a principle protecting fair equality of opportunity because health care ought to protect an individual's fair share of the normal opportunity range. Prudent deliberation about how opportunity must be protected over the lifespan, however, leads to a more specific principle for the design of health-care institutions. They must distribute health care in a way that protects individuals' fair share of the age-relative normal opportunity range for their society. This is the overarching principle that constrains all further deliberation about prudent lifespan allocation of health care.

The Prudential Lifespan Account
and Health-Care Delivery

Our prudent deliberators modify the fair equality of opportunity account by requiring that normal opportunity range be protected at each stage of life. Prudent deliberators using this principle and what else they know about their society can draw some basic conclusions which affect the design of their health-care system. In the remarks that follow, I shall highlight only certain general conclusions implied by the Prudential Lifespan Account. In later chapters I shall discuss some of these issues in greater detail.

One basic conclusion our prudent deliberators draw is that the design of the health-care system must reflect what I shall call the disease/age profile for the society. A prudently designed health-care system will be responsive to facts about the types and frequencies of disease and disability that emerge at different points in the lifespan. For example, in this century, we have seen striking changes in the disease/age profile in industrialized societies. The prevalence of infectious diseases and other actue crises causing early morbidity and death has been dramatically reduced, but the prevalence of chronic disease and disability has just as dramatically increased. Of course, much of the change in the disease/age profile is the result of health-care interventions, both medical and public health, as well as environmental (including nutrition). Moreover, to some extent, the change in disease/age profile will itself have effects on the age profile of a society (the proportion of the population in each age group), though major changes in the age profile have primarily been the result of changing birthrates.

In making the health-care system responsive to the disease/age profile, prudent deliberators will be concerned about a special implication of the age-relative fair equality of opportunity principle they have chosen. The impact on opportunity range of the same disease or disability may be different at different stages of life. As a result, it will be necessary to attribute different importance to meeting those health-care needs at different stages of life. The importance of a particular need is not stable throughout the lifespan.

A second basic conclusion our prudent deliberators draw is that the design of the health-care system must reflect facts about the age profile of the society and about projected demographic changes. The age profile, together with the disease/age profile, when combined with the equal opportunity principle, will determine what *rates* and *types* of transfers of health-care resources are prudent to make between stages of a life—and thus fair to make between age groups.

Prudent deliberators are mindful of an important point about the rate of savings. In general, a longer lifespan will mean a greater rate of savings, where the rate of savings is viewed as a portion of lifetime earnings. If lifetime earnings are held constant, but lifespan is extended, then the rate at which resources must be transfered from early stages of life to late stages must be increased. We must take more from our young and middle years to finance our later ones. Of course, if we remain productive longer, then the savings rate may not have to be increased: But that would be because lifetime earnings have been increased. For example, if we extend average lifespan but encourage people to continue work through their young-old years, then savings may not have to be increased. Wherever lifespan is increased during nonproductive years, however, because we extend the period of old-old age or have more people reach it, then we must save at a greater rate in our productive years. In savings schemes that rely on compacts between successive birth cohorts, increases in productivity may make the increased savings rate required by an aging society less burdensome (I return to this issue in Chapter 7).

Incidentally, a reminder may be in order that our prudent deliberators must draw these conclusions about savings because the age-relative opportunity principle does not let them discount the value of years late in life simply by virtue of their being later. The theory of prudence they begin with requires that they maximize well-being over the lifespan in a manner that is neutral with regard to time. Moreover, they are constrained to think about *themselves* living through each stage of the system they design.

Together, the conclusions our prudent deliberators draw show that some current ways of thinking about "generational inequity" are based on inappropriate views about how to measure equity. In Chapter 1 I referred to critics of current government spending who complain that per capita expenditures on the elderly significantly exceed per capita expenditures for children. Their concern is that these differences, if great enough, may constitute an inequity (it is not clear why these critics actually think that only equal per capita expenditures would constitute equity). But from the perspective of prudent deliberators concerned with meeting health-care needs over the lifespan, for example, per capita expenditures are very likely to differ greatly between age groups. Facts about the disease/age profile and the age profile, plus the emphasis placed by the opportunity account on meeting needs, determine that per capita expenditures are likely to vary greatly between children, the young, and the very old. These variations are what prudence and thus justice require.

Prudent deliberators would have to assess the relative importance of basic types of health-care services in light of the basic conclusions they have already drawn. A crucial decision concerns the importance of long-term-care services—the provision of various levels of nursing-home care as well as home-care and personal-care services needed by the partially disabled elderly. How, for example, would they compare the relative importance of various personal-care and social-support services for the partially disabled with acute-care medical services?

From their perspective, the two types of care would have the same rationale and the same general importance. Personal medical services restore normal functioning and thus have a great impact on an individual's share of the normal opportunity range at each stage of life. But so too do personal-care and social-support services for the partially disabled and frail elderly. These services compensate for losses of normal functioning in ways that enhance individual opportunity. The disabilities that require long-term care are not life-threatening, and people usually live for many years with them. But they can have a dramatic impact on an individual's opportunity to carry out otherwise reasonable parts of a life plan, and quality of life may be sharply reduced if there are no personal-care and social-support services to promote independent living. Since these disabilities affect such a substantial portion of the later stages of life, it is not prudent to design a system that ignores them and meets only the acute-care crises of the elderly. This form of imprudence involves either not transfering enough resources to the later stages of life—too low a rate of savings—or transfering resources in the wrong form (making claims on acute services instead of personal-care and social-support services).

About 80 percent of long-term care in our society is provided by families to their elderly relatives. Prudent deliberators would be very careful to design their long-term-care system so that it is responsive to the needs of both the elderly and the families providing such care. For example, a system of day care centers that would relieve families of some of the burden of around-the-clock care would benefit the elderly, who greatly fear burdening their children and grandchildren. This system would also benefit the adult children who need respite from the burden of providing care because of career demands or other family obligations. I return to these very important conclusions about long-term care in Chapter 6.

The Prudential Lifespan Account also has implications for the relative importance of other categories of health-care services, for example, preventive versus curative services. If we were to design a health-care system from a perspective that emphasized the conflicting interests between different age groups—indeed, between different birth cohorts—then we would have to be concerned about the fact that the current young will benefit from the provision of preventive health measures, whereas the current elderly will benefit little from them. Prudent deliberators are concerned with the relative importance of health measures which operate over the whole lifespan, however. Thus they would not discount the importance of preventive measures.

Since resource constraints may vary from society to society, prudent deliberators will draw different conclusions about rationing for different systems. In general, their decisions reflect the ways in which services will have an impact on the age-relative normal opportunity range, which is their crude measure of the importance of meeting different health-care needs. The provision of one service, old or new, high or low technology, carries with it opportunity costs. When choices are made to provide one set of services, other needs are not met, and these unmet needs count as foregone opportunities or ''opportunity costs.'' All these costs are thought of as contained within the system. The justification for meeting one set of needs rather than another is that it is more important to do so, in view of effects on opportunity.

This general form of rationing does not involve appeals to *age* as a criterion of distribution. In general, the fair equality of opportunity account focuses directly on comparisons of need for its rationing decisions. The possibility remains, however, that under some conditions of resource scarcity, it may be prudent to ration certain resources by age, where the appeal to age is not a disguised way of referring to medical suitability or to need. For example, there is considerable evidence that the British National Health Service rations dialysis and some other technologies by age. To the extent that such rationing is not based on (erroneous) views that the elderly are not medically suitable for dialysis (cf. Aaron and Schwartz 1984), then the rationing is purely by reference to age. This form of rationing needs a justification. In Chapter 5 I shall consider arguments about the moral permissibility of rationing life-extending health-care resources by age. It is an implication of the Prudential Lifespan Account that such rationing cannot be ruled out as morally impermissible under all conditions.

The prudent design of a health-care system governed by the equal opportunity principle would have significant implications for other aspects of health planning. If the system must emphasize primary-care and personal-care services for the very old, for example, that has implications for manpower. It might mean we must produce more geriatricians and trained providers of personal-care services. Similarly, there will be implications for priorities in research that derive from judgments about the importance and feasibility of meeting certain health-care needs. If there are social obligations to provide a particular array of services because these constitute a prudently designed system that protects equal opportunity, then we must provide the incentives and resources needed to attract and train personnel who deliver the care and carry out the research that makes it feasible. I shall say little about these features of institutional design in what follows, but there is no reason to believe that the requirements of justice in this regard involve violating any important liberties of health-care providers (cf. Daniels 1985, Chapter 6). Nor would rationing in the system mean that providers must violate duties to their patients (on this point see Chapter 8).

Now that we have seen some of the implications of the Prudential Lifespan Account, it is possible to clarify a remark made earlier about rights to health care. If health-care rights are a species of rights to equal opportunity, as I have argued earlier, then the actual entitlements such rights give rise to will depend on the prudent design of the health-care system. Specifically, our health-care rights might give us legitimate claims to services at one stage of our life but not at another. This may happen because meeting certain needs is more important at one stage of life than at another, or it may happen because life as a whole will be better if resources are rationed by age. The inequalities in entitlements held by different age groups do not, however, mean that people are being treated unequally, at least over the course of their lives, as I pointed out earlier. Over the lifespan, our rights to health care will be *equal* rights, even if those equal rights yield unequal entitlements at different points in the lifespan.

Summary

My task in this chapter was to fill out details of the Prudential Lifespan Account as an approach to the problem of justice between age groups. Specifically, I have considered how prudent deliberators, behind the appropriate veil of ignorance, would think about the allocation of fair shares of health care over their lifespan. The notion of a fair share for health care is determined by the fair equality of opportunity account, but prudent deliberators must modify or adapt that general principle to the problem of allocation over the lifespan. In this case, they choose as their fundamental principle governing the design of health-care institutions a principle that guarantees individuals a fair share of the normal opportunity range for their society *at each stage of life*. Once this principle is chosen, information about the disease/age profile and the age profile for their society will lead to specific conclusions about what kinds of health-care services are most important to provide at each stage of life. Designers of the system must also make choices about the relative importance of basic categories of services—such as acute versus preventive versus long-term-care services. And they will have to ration these services, depending on the stringency of constraints on resources, in accordance with the fair equality of opportunity principle.

Two specific implications of the Prudential Lifespan Account will be developed in subsequent chapters. In Chapter 5 I consider the moral permissibility of rationing health-care resources by age. In that discussion, I shall also examine an issue postponed from Chapter 3, namely, the "moral relevance" of age as a distributive criterion. In Chapter 6 I take up the argument sketched here that long-term-care services are much neglected in the United States and that the just design of a health-care system would require adequate provision of them.

I shall return to the problem of income-support policy in Chapter 7, where I shall discuss some implications of the Prudential Lifespan Account on the distribution of income between age groups. A central concern of Chapter 7, however, will be the relation between the age-group and birth-cohort problems. Much of the recent discussion of the stability of our Social Security and Medicare systems in the face of the aging of society concerns this issue.

5

Rationing by Age

Rationing by Age in
the British National Health Service

Is it morally permissible, from the perspective of justice, to ration life-extending resources by age? I do not have in mind rationing in which age stands as a surrogate for, or rough indicator of, some other property of persons, such as medical suitability. Rather, I want to consider what might be called *pure age-rationing.*[1] The question is not purely hypothetical, however. Although it might be politically infeasible to advocate pure age-rationing in the United States, the combination of an aging population and rapidly growing health-care costs may soon require us to face the issue. Moreover, there already is age-rationing, though it is not clear how "pure" it is, in the British National Health Service (BNHS).

In their important book on rationing hospital care, Aaron and Schwartz (1984) document the way in which hemodialysis is rationed

---

1. Allen Buchanan has pointed out to me that the rationing by age which I end up defending is not pure in the following sense: There is a reason for appealing to age which has to do with effectively promoting opportunity. Still, age is not a surrogate for other traits, like medical suitability, and that is the sense in which rationing by age is pure.

by age in the BNHS. Few people beyond the 55–65 age bracket are given dialysis—and those who get it probably have made a fuss in order to circumvent the normal policy. There is evidence that British physicians think elderly patients are not medically suitable for dialysis, that they are "crumbly" (Aaron and Schwartz 1984:35). Most American physicians would probably dispute this assessment and would view the British judgment about medical suitability as medically unfounded. Let us assume that for many British practitioners, the judgment is sincere and not a conscious rationalization. (If it is an insincere rationalization, we have a serious case of deception, which raises other moral issues.) On this assumption, not recommending the elderly for dialysis is not a case of pure age-rationing, as I have defined it. Still, there is some reason to think that the British practice, taken at a policy level, does involve a resource-allocation decision that comes as a form of pure age-rationing. I shall return later in this chapter to some features of actual BNHS practice that I believe violate some requirements of justice. First, however, I want to develop an argument about pure age-rationing in theory before I discuss it in practice.

The reaction of most who hear about the BNHS policy regarding dialysis is to condemn it as "ageist." Critics view it as morally unacceptable discrimination, in the same way rationing by sex or race would be. In what follows, I shall defend a contrary view, offering an argument which I believe shows that pure age-rationing is morally permissible under certain, very specific, and restrictive conditions. Such an argument could possibly provide a rationale for the BNHS policy (though as I already noted, that policy may be faulted on other grounds). My point, however, is not to defend the British policy or to advocate anything like it in the United States. Rather, I want to use the Prudential Lifespan Account to clarify further when a distributive policy or practice is age-biased in a morally objectionable way. Pure age-rationing provides us with a limiting case that can illuminate the more general problem.

Judging from the outrage many express at the British for age-rationing dialysis, one might think we in the United States had a consistent policy prohibiting appeal to age as a distributive criterion. Legislation prohibiting discrimination on the basis of age in educational, employment, and housing contexts reinforces this impression. But the picture is more complex. Many policies and practices in the United States do involve an appeal to age as a distributive criterion. Relatively uncontroversial are the uses of age criteria to stand as rough indicators of competency. We let only those over age 18 vote or serve in the armed forces, those over age 16 drive, and those over 18–21 (depending on the state) drink alcohol. At the other end of the lifespan, senior citizens receive discounts on public transportation, movies, tourist attractions, and many other privately marketed items. Public programs appeal to age criteria for the distribution of economic and medical benefits—there are age restrictions on eligibility of children for AFDC benefits and on the eligibility of adults for retirement or Medicare benefits. Moreover, public programs with age-eligibility requirements are specifically exempt from provisions of the laws prohibiting discrimination by age.

When is the use of an age criterion a legitimate aspect of public policy, and when is it morally objectionable? What is the "moral relevance" of age in distributive contexts? I postponed consideration of this issue in Chapter 3 but I must now address it. In what follows, I shall first appeal to the Prudential Lifespan Account to develop an argument about the permissibility of rationing by age under some conditions. I shall then distinguish this argument from other kinds of arguments—both pro and con—with which it might be confused. After some further discussion of British rationing by age, I shall return to the more general question I have raised: Is there a principled and morally acceptable way to make sense out of appeals to age as a criterion in distributive contexts?

A Prudential Argument
for Pure Age-rationing

The Prudential Lifespan Account is concerned, we recall, with institutions that distribute basic goods over the lifespan. By finding out what rational deliberators, operating under certain information constraints, would accept as prudent to allocate to different stages of their lives, we also discover what is fair between age groups. These deliberators work within a *frame* that limits the scope of their problem. They are to allocate fair shares of health care, or other basic goods, which means they are not trying to solve problems of distributive justice that involve transfers of goods across the boundaries between persons. These deliberators must also assume they will live through each stage of life under the institutions they design. Finally, they do not know their own individual plan of life, a restriction that prevents the scheme from being biased toward one stage of life. The details and justifications for these constraints have been described in Chapters 3 and 4.

The crux of the argument is easy to state and was alluded to in Chapter 4. There we noted a relationship between measures that affect longevity and rates of saving. We need to reconsider this before I go on to explore my next claim, which will be that under certain resource and information constraints, prudent deliberators would prefer a distributive scheme that improves their chances of reaching a normal lifespan to one that gives them a reduced chance of reaching a normal lifespan but a greater chance to live an extended span once the normal span is reached.

Let's look more closely at the relationship noted before. If lifetime earnings are held constant, but lifespan itself is extended, then the rate at which resources must be transferred from early stages of life to later stages must be increased. We must take more from our young and middle years to finance our later ones. Of course, if we remain productive when we are old because we work longer, then the savings rate may not have to be increased. Wherever lifespan is increased by extending nonproductive years, then we must save at a greater rate in our productive years to cover the expenses of the extra nonproductive ones. Here, of course, prudence is at work: It would be imprudent not to provide resources for the later stages of life when we have reasonably good chances of having to live through them. Similarly, where increased longevity is primarily achieved by reducing early death, for example, by measures that reduce infant or young adult mortality, the increased productivity that results will (we may suppose) roughly counterbalance the need to save more. Where increased longevity results from marginally extending the lives of the very old (especially those unable to work), however, then savings will have to be increased.

Under some resource constraints, the increased rate of savings needed to provide prudently for a lifespan extended beyond the normal range will have serious negative effects on early stages of life. We can imagine constraints which operate in the following way: Providing very expensive or very scarce life-extending services to those who have reached normal lifespan can be accomplished only by reducing access by the young to those resources. Saving these resources by giving ourselves claim to them in our old age is possible only if we give ourselves reduced access to them in earlier stages of life. A central effect of this form of saving is that we increase our chance of living a longer-than-normal lifespan at the cost of reducing our chances of reaching a normal lifespan.

We see a version of this effect in recent health-care policy. Infant mortality rates in Massachusetts stopped decreasing for a period after 1980. This trend coincided with drastic cutbacks in prenatal care, especially for the poor (Knox 1984). In the meantime, despite recent cost-containment measures, we leave untouched the acute-care bias of the Medicare system. We continue to devote vast hospital resources to marginally prolonging the life of the dying elderly, who are a substantial portion of "high-cost users" (Zook and Moore 1980). Public funds thus have been used in a way that reduces slightly the chances of reaching a normal lifespan but increases the chances of living an extended lifespan for those who reach normal life-expectancy.

Let's make this point more concrete by considering two rationing schemes. Scheme A (Age-rationing) involves a direct appeal to an age criterion: No one over age 70 or 75—taken to represent normal lifespan—is eligible to receive any of several high-cost, life-extending technologies such as dialysis, transplant surgery, or extensive by-pass surgery. Because age-rationing reduces utilization of each technology, there are resources available for developing them all, though under this scenario that development will be only for the young. Scheme L (Lottery) rejects age-rationing and allocates life-extending technology solely by medical need. As a result, it can either develop just one such major technology, say dialysis, making it available to anyone who needs it, or it can develop several technologies, but then ration them by lottery.

Scheme A saves resources—defers their use until later in life—at a lower rate than Scheme L. Scheme L takes more from earlier stages so that later ones may benefit. Specifically, Scheme L involves reducing the chance that the young will reach a normal lifespan because access to life-extending resources has been reduced. In return, Scheme L offers an increased chance of living a longer than normal span to those who do reach normal lifespan. For instance, though this is an extreme example, Scheme A might offer a 1.0 probability of reaching age 75 (and dying right away), and Scheme L might give a .5 probability of reaching 50 and a .5 probability of reaching 100. Both yield the same expected lifespan, but they do so differently. (Intuitions about science fiction cases are always of questionable utility, but for those who insist: Imagine there is a disease around which would kill everyone at age 50, but a drug is available in short supply. We can give a half-dose to everyone, and they will then live to 75 and die. Or we can give a full dose to half the population by lottery at age 50; lottery winners will live to age 100 and die right away, but lottery losers will die right away at age 50. The two scenarios produce equal average life-expectancies.)

Our prudent deliberators must choose between Schemes A and L (leave aside the science fiction example). I shall argue that prudent deliberators would probably prefer an age-rationing scheme to a lottery. The arument is complicated, however, by the vague way in which I have described conditions under which deliberators must choose, as anyone familiar with recent work in the theory of justice will have noted. I have left the description of these conditions vague because defending a particular construction is difficult and, for our purposes here, digressive. But the price I pay for sticking to the point is that I must now consider the way the argument might run under alternative constraints. Specifically, we must consider two alternative rules of rational choice that might be invoked to govern prudential reasoning about Schemes A and L.

One rule is the "maximin" rule (maximize the minimum), which tells us to make the worst outcome as good as it is possible to make it (which might mean to make it as unlikely to happen as possible). Maximin is appropriate to governing rational choices when real uncertainty—not just risk or probability—is a feature of the choice situation. That is, we cannot invoke likelihoods of outcomes, or even reasonably assume outcomes are equi-probable and assign numbers to them. Some might claim that maximin is also the appropriate rule when the worst outcomes are so grave that they cannot merely be weighed against better outcomes.

I have not described the choice as one in which the maximum rule is clearly the appropriate one. If it is the correct rule, then I think it is easy to show that Scheme A would be preferable to Scheme L. Suppose someone is inclined to like Scheme L because it seems to help us most later in life. This person thinks the life of the wise, revered elder is the best thing one would aim for, and does not mind taking some chances (if one could calculate them) to reach the Golden Age. Even such a person would have to admit that the worst outcome would be dying young. If we can assign no probabilities in our reasoning, for example, we can assign no probabilities to whether we are likely to die young or to live to the Golden Age, we are constrained by the maximum rule to minimize the likelihood of the worst outcome. This would force us to choose Scheme A.

I have not insisted on the maximum rule because I am not sure I want to insist that the "veil of ignorance" surrounding our prudent deliberators be so thick that they cannot predict the relevant estimates of the likelihood of longevity under the two schemes. After all, I want our deliberators to know enough about their social system so that they can make prudent judgments about the design of its health-care system. This means they must know something about its specific demography and the way it is affected by alternative arrangements of the health-care system. As a result, we might think it reasonable to adopt a more common rule of choice. This Standard Rule, as I shall call it, instructs prudent deliberators to maximize their expected net benefit or payoff when they face choices. It requires that they take into account not only the value of a payoff, but its likelihood or probability, and that they maximize the product of the two.

How would Schemes A and L fare under the Standard Rule? Suppose, for the moment, that we take as the payoff the number of years lived. The Standard Rule tells us to maximize the expected lifespan. If our choice between Schemes A and L is a choice between schemes that each give equivalent expected lifespans—for example, the choice between a 1.0 probability of living to (and only to) 75 under Scheme A and, under Scheme L, a .5 probability of living only until 50 and .5 probability of living (only) to 100—then the Standard Rule instructs the prudent deliberators to be indifferent. In the absence of a better scheme, there is a tie: Each must be deemed prudent and both are acceptable.

Even this tie is an interesting and important result. It tells us that age-rationing cannot in general be ruled out on the grounds that it is imprudent. This means in turn that it cannot be ruled out under the conditions I have argued are appropriate for deciding what is just between age groups. Thus we cannot claim that age-rationing is always unjust.

The argument I have just sketched for the Standard Rule seems too abstract, even for the situation I have described in which parties do not have knowledge of their conceptions of what is good in life. Even without the details of such knowledge, we still know enough about the frequencies of disease and disability as we age to know that years late in life, say after age 75, are far more likely than earlier years to involve some forms of impairment. This knowledge suggests that it would be imprudent to count the expected payoff of years late in life quite as highly as the expected payoff of years more likely to be free of physical and mental impairment. To be sure, many people enjoy their later years relatively free of impairment—I am not drawing on a stereotype that all the old are frail and sick—and there is no suggestion here that age by itself gives us any basis for judging the value of these years less. Moreover, many people with impairments would admit to being no less happy than other people without impairments. Some people are happy and cope well, though others do not. Nevertheless, the prudent deliberators are estimating expected payoffs, which means they should take into account the frequencies of disability and disease. They then should discount the expected payoff of later years accordingly. Consequently, they would reject the idea that there is a tie between Schemes A and L, merely because life expectancies are equal. After considering disabilities, Scheme A would again seem more prudent.

We can think about the choice between Schemes A and L under the Standard Rule in a slightly different way, which may be an alternative or supplement to the above argument. Under some plans of life, the contribution of the last years to the overall meaningfulness of life might be very great. Still, such Golden Age plans are probably atypical. Most people are well aware of their mortality and construct plans in which the tasks and rewards of early and middle years are integral to their success. For them, later years can be wonderful, but they are gravy to the meat and potatoes of the rest of life. Without making the judgment that one plan of life is better than another or even, by itself, less prudent, deliberators, familiar with their society and culture but unaware of which conception of the good is theirs, would estimate it to be more likely that they will have typical plans than Golden Age plans. They might then select Scheme A over Scheme L because they want to increase their chances that they live through the middle stages of their lives: That is what will most ensure success of their probable plan of life.[2] Notice that if we want to block completely this type of proba-

---

2. Allen Buchanan has suggested to me that there is another important reason why rationally prudent individuals would discount later years. *Fecundity* of benefits is one important factor to take into account in calculating expected benefit. But if this is so, then, other things being equal, a later year is worth less than an earlier one because whatever opportunities for generating further benefits from activities pursued in a later year are less than those generated at an earlier year. The same reasoning is what leads economists to discount future dollars; they are worth less because there is less time to invest them and

bilistic reasoning, then we start to push ourselves into a description of the problem of choice that makes the maximum rule seem more appropriate than the Standard Rule. Then the deliberators would choose Scheme A in any case.

My conclusion from these versions of the prudential argument is that there are conditions under which a health-care system that rationed life-extending resources by age would be the prudent choice and therefore the choice that constituted a just or fair distribution of resources between age groups. Since the argument I have offered leads to such a controversial conclusion, and since it is a complex argument with many presuppositions, I would like to distinguish it explicitly from some kinds of arguments—for and against similar conclusions—with which it may be confused.

Nonprudential Arguments For and Against Age-rationing

The prudential argument I have offered focuses on rationing by age, but it has broader implications, in particular, for discussions about the relative merits of extending the biological lifespan versus improving functionality within the normal span.[3] The topic has thus attracted some discussion from diverse points of view. I want to be careful to distinguish the argument I offer from others, both for and against my conclusion, because I think what is distinctive about my argument avoids much that is objectionable in these other arguments. In the course of these remarks, I shall perforce reply to some objections to my argument.

Consider first an anti-age-rationing argument that appeals to equality. An age-rationing scheme will not treat the 75-year-old the same way it treats a 55-year-old, even though there may be no difference in their medical condition or need. But this is unequal treatment of like cases, for age is not a "morality relevant" consideration, but medical condition is. So any scheme that rations by age is morally objectionable.

This argument really makes two distinct claims, one about equality

---

hence they are less fecund, a point which has nothing to do with inflation.

3. This is referred to as "squaring the curve"; cf. Vaupal (1976) and the essays in Veatch (1979).

and the other about moral relevance. The scheme I have justified by my prudential argument is not open to the objection about equality, at least if we treat the notion of equality carefully and we observe how the schemes we choose operate over a lifetime. This point, it will be recalled, was the focus of discussion in Chapter 3, pp. 40–42. Specifically, the 75-year-old and the 55-year-old will not be treated differently over the course of their lives. Before each is 75, he will be entitled to the life-extending treatment in question; afterwards, he will not. Thus treating someone now 75 differently from someone now 55 does not constitute unequal treatment of the whole person over his lifetime. (Remember, we are ignoring the complication that arises when we introduce an age-rationing scheme in the middle of someone's life; see Chapter 7.)

The issue of moral relevance was postponed in Chapter 3, but we can now begin to consider it. To say that age is a morally irrelevant feature of persons is question-begging. From the perspective of prudential planning over a lifetime, which I have argued is the appropriate perspective to adopt, it may be quite relevant. The prudential argument I have offered centers on the claim that considering age can produce an allocation scheme that makes the life of *each person better off* (at least, not worse off) as a whole. Since whatever is prudent from this perspective constitutes what is just, it is at least morally permissible and may be morally required. If I am right that the Prudential Lifespan Account is the appropriate method for solving the age-group problem, then we cannot decide a priori that age is a morally irrelevant criterion. Moral relevance can be determined only by seeing what distributive schemes it is prudent for deliberators to select. I shall say more about this issue of moral relevance later in this chapter.

Consider now a pro-age-rationing argument which appeals to equality or fairness in a different way. We should give priority to the young in rationing life-extending services, so the argument goes, because the old have already had a chance to live more years and it is only "fair" to the young to give them an equal chance.[4] This appeal to intuitions about fairness is not persuasive. Does it matter to our intuitions whether the old have already made claims on comparable resources to extend their

4. Veatch (1979) offers a variant on this argument in which the principle requiring the equal chance is itself derived from a contractarian argument intended to yield a general principle of justice for health care. I implicitly reject Veatch's argument in Daniels (1979, 1981, and 1985).

lives, or is the occasion of competition with the young their first such claim? What if the young person has already received a lot of help, but the old person none? Is it still fair to the old to deny them help now? When do we say the young have used up their "equal chance" to live to an old age?

I am not sure where such intuitions lead us or whether they are to be trusted. My prudential argument is different in that it appeals to no prior intuitions about fairness, except perhaps those which require that the choice be procedurally fair to the prudent deliberators. The fairness of the outcome—what counts as just is what prudence recommends under these choice conditions—is derived from the fairness of the choice and the fact that it is reasonable for us to let these deliberators decide for us (cf. Rawls 1971, 1980).

One familiar type of argument against age-rationing appears to rest on a fundamental claim about the equal value of life at any age and seems to deny part of the argument I offer above. Thus some insist that age-rationing is wrong because "life at any age is worth living." Others insist that "life at any age is of equal worth to life at any other." This claim about the value of life is already accommodated within my prudential argument.

From the perspective of prudent designers of a health-care scheme using the Standard Rule, the most the equal worth claim can imply is that they should choose schemes which maximize expected lifespan, that is, what valuing each year of life equally would amount to. By itself, however, this appeal to equal worth does not decide the question for or against age-rationing. In particular, deliberators would have to consider the age-rationing and lottery schemes equally prudent if they had the same (maximizing) effect on expected lifespan. They would then have to be indifferent in the choice between them. Thus the equal value assumption does not of necessity lead to opposition to age-rationing. Moreover, if a lottery scheme failed to maximize life-expectancy as well as an age-rationing scheme, then the commitment to "life at any age is worth living" would in fact commit one to choosing the age-rationing scheme. The assertion of equal worth should not be understood only from the perspective of the elderly who know they have already lived many years and want more; it should be interpreted from the perspective of the prudent deliberator seeking a system that maximizes the expected lifespan. Then age-rationing is not ruled out.

Sometimes this "equal worth" argument is coupled with the complaint that no one other than the person whose life is in question should judge its worth. This complaint seems to strengthen the anti-age-rationing argument because we may suspect that the rationing is being advocated by the young, who have an interest in it, and therefore any judgment that seems to deny equal worth is morally suspect. It is an important feature of the prudential argument I offer, however, that it involves no judgments by one person about the value or worth of another's life. Instead, it involves persons making judgments *for themselves* about benefits *to themselves* at different stages of their lives.

In this regard, the prudential argument works differently from the age-rationing argument that claims ("their") elderly lives are not worth what ("our") younger ones are. Examples of this type of argument are not hard to find, though sometimes the claim is only implicit. A version of the claim is embedded in certain methods for "pricing" life. It is explicit in an earnings-streams approach, which values a year of life proportionately to the earnings that flow from saving it. Younger years are thus worth more than older years. The same valuation will be implicit in some willingness-to-pay methods, which ask people how much they would be willing to pay to save a year early in life rather than a later one. Alternatively, some urge that it is more cost-beneficial or cost-effective to use certain resources to save the young rather than the old because of the difference in "quality adjusted life years" saved. With all of these approaches, there is a better return on the social investment of resources if the young get treated and the old do not.

In the prudential argument I offered earlier, where reference is made to the increased frequency of impairments in later life or to the likelihood that an earlier year will be more important to a plan of life than a later year, the argument may sound similar. But the prudential argument involves reasoning about the expected benefit of an extra year at two points in one life—the life of the prudent deliberator. This judgment is not one "we" make about "them" but one whose consequences we must live with in each stage of our lives. Prudent deliberators draw the conclusion that their own lives are less important to extend at one stage than at another—and that is the kind of judgment we do in fact make when we prudently plan our actual lives.

The prudential argument, especially under the Standard Rule, has some resemblance to a more general form of utilitarian reasoning even though it is not an instance of such an account. There are two important contrasts between my argument and a utilitarian one. First, the prudential argument takes place within the frame provided by the fair equality of opportunity account of just health care. In turn, though the equal opportunity principle may be compatible with some forms of ideal rule utilitarianism (cf. Brandt 1979), it is established independently of any appeal to utilitarianism. The frame around my prudential argument means that I am not using this form of prudential reasoning as an account of just health care or justice in general. Thus the utilitarian appearance of the argument is misleading. It is easy, however, to lose sight of the frame provided by the equal opportunity account in the argument rationing life-extending resources by age. The frame seems to play no direct rule in the argument because the extension of life itself has a comparable effect on the age-relative opportunity range at any stage of life. Consequently, impairment of the age-relative normal opportunity range would not decide the particular rationing question we were discussing, even if it does have a bearing on other allocation issues between age groups.

There is a more interesting contrast as well. The utilitarian who emphasizes "quality adjusted life years saved" is interested in showing that a policy maximizes *overall* utility. In contrast, the prudential argument says that age-rationing would be acceptable when it does better for *each of us* to budget resources over the lifespan in a certain way. The prudential argument thus explains the general intuition that the British system can be defended on the grounds that it "does more good": It does so for each of us. Although the utilitarian account of doing more good in the aggregate allows us to take from some people to give to others, the prudential account avoids crossing the boundaries between persons in that objectionable way.[5]

5. I am indebted to Allen Buchanan for the above point. He has also suggested to me that *if* my view were correctly implemented, then the change from what we now have in the United States to what would result would not only be a Pareto Improvement, it would be a Strong Pareto Improvement. That is, everybody would be better off, not just some better off and no one worse off, in terms of what benefits are gained from the fair share of health-care resources. Finally, even if my view is not utilitarian, it should be attractive to a utilitarian, for though it does not maximize utility, it tends to increase it. Thus the Prudential Lifespan Account is attractive to both the utilitarian and his chief rival, the Rawlsian, who wants us to use resources in a way that effectively promotes fair equality of opportunity for each of us.

Remarks about the
British National Health Service

It is easy to misconstrue and misapply my argument. The prudential argument does not, in general, sanction rationing by age. Such justification is possible only under very special circumstances. First, it is crucial to establish that the appeal to an age criterion is part of the design of a basic institution that distributes resources over the lifetime of the individuals it affects. Nothing in the argument offered here justifies the piecemeal use of age criteria in various individual or group settings— for example, by some hospitals or physicians, or in any way that is not part of an overall prudent allocation.

Second, the argument should not be taken as a hasty endorsement of age-rationing as a convenient "cost constraining" device in the context of current debates about our health-care system. Not only is such an application not likely to be part of the design of our basic health-care institutions, construed as a savings scheme, but many of the assumptions about resource scarcity, which might make rationing by age prudent in some circumstances, are controversial in the context of this public policy debate. This does not mean that my prudential perspective cannot be suggestive of some reforms we might aim for, for example, more extensive use of advance directives through which people give instructions about what medical interventions they want under certain contingencies (see Chapter 8).

Finally, it is important to see that my argument is part of an ideal theory of justice, in which we can assume general compliance with principles of justice which govern other aspects of our basic social institutions. The argument does not readily or easily extend to nonideal contexts, in which no such compliance with general principles of justice obtains. Thus it would be wrong to say that my argument would actually justify the British system of rationing dialysis by age (if that is in fact their practice). At most my argument shows that such rationing can under some circumstances be part of a just institution, that it is not always morally objectionable in the way that rationing by sex or race would be. The argument shows the conditions that would have to obtain for such rationing to be just.

Suppose, however, that this last qualification is unnecessary and that we should accept the view that the British health-care system is in general a just one. Let us also suppose that at some level in the BNHS, the age-rationing decision is a pure one and not based on sincere belief in the medical unsuitability of elderly patients for dialysis. Does my argument about pure age-rationing constitute a defense of their practice? I believe that this British practice may be faulted on other grounds of justice, but not on the grounds that age-rationing is always unacceptable. Rather, the problem is that the British scheme depends in part on the lack of public awareness of the basis for the rationing of dialysis by age.

What happens in practice is that a patient or the family will be told that "there is nothing more than can be done for you." This may be the message delivered by the British general practitioner, who will then not refer the patient to the specialist (who alone can order dialysis), or it might be the message of the specialist himself. The patient most likely will assume there is nothing that can medically be done, but in fact, underlying the denial of treatment, is a rationing decision to conserve scarce dialysis resources for younger patients. If the physician is suspicious that this is a pure age-rationing decision, and that medical suitability is really not the issue, then there is a deception in which he shares directly. Even if he sincerely believes in the medical unsuitability of the elderly, then there is, on the supposition I am making, a deception, though now the physician is just an instrument through which the deception is carried out. If the rationing by age is pure, then the grounds for denying treatment must be that there is nothing that society *is willing* to do for the patient because of his or her age, not that there is nothing more that can medically be done.

An interesting question now arises. Suppose that the only way for an age-rationing scheme to remain stable is for the grounds for the rationing to be disguised, and yet there are good prudential reasons for implementing such a scheme. Is the scheme just? In general, I think that principles of justice and the reasoning for them must be public, and that the conditions of publicity are rather stringent. (Rawls [1980] argues for such strong publicity conditions for the general theory of justice.) This does not mean, of course, that the whole public must agree with or accept the scheme. Nor does it mean that proposed policy is not a just one if it cannot be made acceptable to a particular society: It may be that what is ideally just is not feasible under some conditions, that is, if the publicity constraints are adhered to. We then need an account of what is permissible when what ought to be justifiable is politically infeasible. Here philosophers have generally fallen silent, and I shall too.

Age and Moral Relevance

The prudential lifespan argument in this chapter establishes one narrow result: Under very special conditions of resource scarcity, it is not unjust to ration life-extending health-care resources purely by age. This result stands in sharp contrast with what is probably the dominant view in the United States about the age-rationing of dialysis in Great Britain. The dominant view reflects the principle that health status, not age, is the morally relevant basis for distributing health-care resources. The prudential lifespan argument shows that this principle cannot be maintained in this strong form.

I noted early in this chapter that our moral and legal practice is not in general clear about the moral relevance of age as a criterion in distributive contexts. We use age as an eligibility criterion in many government programs that distribute basic resources, as well as in fragmented public and private practices that grant special privileges or discounts to certain age groups. At the same time, we prohibit employment practices that award jobs or promotions on the basis of age, and we prohibit discrimination by age in other contexts as well. Does the Prudential Lifespan Account tell us anything more general about the moral relevance of age in distributive contexts? These questions deserve at least a brief answer.

It is important to understand that moral relevance is not a *formal* but a *substantive* moral notion. We cannot, for example, tell by mere inspection of the concepts involved whether a trait is morally relevant because it stands in a formal or logical relationship to some other concept. To judge the moral relevance of a trait of individuals for some distributive purpose, we must know what is being distributed and we must have some *substantive moral* theory or account of what the appropriate basis for distribution is.

In contexts of employment, for example, race, sex, and religion are in general morally irrelevant. The social good being distributed—jobs or job opportunities—is one we generally think should be distributed in ways that match the requirements of a job to the talents and skills of applicants. This is waht is meant by *meritocratic* placement (cf. Daniels 1978). In general, society has an interest in the enhanced productivity that results from good matches. We may even think we ought to redistribute some of the benefits of such good matches to those who are worst off with respect to talents and skills and who tend to get the worst jobs under meritocratic arrangements. This is the sense in which Rawls views talents and skills as a social asset (cf. Rawls 1971). Nevertheless, however we restrict the inequalities that result from meritocratic job placement, the central principle is that only talents and skills count as morally relevant traits for job placement.

This principle seems to underlie our antidiscriminatory legislation, which prohibits awarding jobs on the basis of sex or race. Even schemes that try to establish better representation of races and sexes in certain jobs through quotas are criticized by some as violating this principle. Exceptions are allowed, however, for special contexts in which sex, for example, is a "bona fide occupational qualification" or a matter of "business necessity."[6] There are also borderline cases in which there is real controversy about what counts as a relevant talent or skill. For example, some workers are specially sensitive to workplace toxins. Does their lack of normal "durability" on the job constitute a lack of a skill or talent, similar to a lack of manual dexterity? I believe the underlying theory leaves this question unresolved (see Daniels 1985, Chapter 8).

In general, we tend to view age as a morally irrelevant trait with regard to job placement. Despite extensive negative stereotypes, which function much like race or sex stereotypes, there is no good evidence that age can stand proxy for job performance in the general range of jobs and offices. Of course, there may be jobs that involve, for example, physical performance at a level that only the very young can achieve. Such special exceptions aside, age is not a good predictor of job performance.

Just as age is not in general morally relevant to job placement, it is also not in general the morally relevant basis for distributing health-care resources. Rather, health-care needs are. Still, we found circumstances in which age became morally relevant to health-care rationing according to the Prudential Lifespan Account. Is it possible that age might also be morally relevant to job placement under some circumstances?

I think the answer is "yes," at least in theory. Age would count as morally relevant provided we could construct a prudential argument analogous to the one for rationing life-extending resources by age. The

---

6. Cf. Daniels (1985, Chapter 8) for discussion of these exceptions.

analogous case would involve restricting access of the elderly to jobs, say through compulsory retirement, to improve the opportunity for younger workers to obtain employment. Of course, the argument requires a frame which is provided by more general principles of justice. Such principles might regulate allowable lifetime inequalities in income and wealth, and guarantee fair equality of opportunity with regard to jobs and careers. It is conceivable that under some conditions prudent deliberators would find such age-rationing of jobs prudent—even taking into account its implications for rates of savings over the lifespan. Whether age-rationing in such an economy would be a better measure to use than other types of economic reform depends on other facts in the hypothetical case, which I will not pursue further. Indeed, I draw only the rather weak conclusion that such cases are not inconceivable.

The general point is that age can become morally relevant to distribution in special circumstances even if it is not ordinarily relevant to the distribution of the basic good in question. That this should be true is no more surprising—or less disconcerting—in employment than it is in health care. In both, the special circumstances lead to a qualification of the general principle that age is not morally relevant to distributing either health care or jobs. The Prudential Lifespan Account is a device that allows us to isolate those special cases while still preserving the (now qualified) general principle.

Does the Prudential Lifespan Account help clarify the diverse other contexts in which we sometimes accept and sometimes reject age criteria? In general, it says that programs which involve age-entitlements—such as Social Security or Medicare—are morally permissible when they are part of the design of a prudent savings scheme. Such age-entitlements in these cases are part of a *systematic* savings-scheme. It is not clear that the diverse special discounts and privileges many elderly receive are part of any such prudent savings-scheme; they seem less defensible. Perhaps some such measures are honorific, a way of showing respect. Still, the majority of them seem to be only promotional efforts to appeal to a specific age group—measures that would be clearly unacceptable if we were taking about race or sex. And some such measures may serve only to build negative stereotypes, for example, that all the elderly are needy. No clear sense can be made of many of them from the prudential lifespan perspective.

A word needs to be said about the debate among many advocates for the elderly that focuses on the problem of "age versus need" (cf. Neugarten 1982). Some advocates for the elderly have expressed concern that age-based entitlement programs may be ill-conceived in either of two ways. They may be difficult to reconcile with the general prescription that age is not a morally relevant basis for distributing basic resources. And they may lead to poor targeting of such resources within redistributive programs. The most needy elderly may not get adequate income support or health-care services, for example, whereas the well-off elderly benefit from programs they do not really need.

The Prudential Lifespan Account can help clarify the issues involved in both of these concerns. First, if an age-based program is part of a prudent savings scheme, then the apparent conflict between using age as a basis for distribution and wanting anti-age-discrimination legislation in other contexts disappears. Age is morally relevant in some distributive contexts but not others. There is no universal, in-principle objection to using age as a basis for distribution. The Prudential Lifespan Account helps us decide when it is morally relevant and when it is not. Moreover, the general difference between age and sex or race remains. Differential treatment by age, unlike differential treatment by race or sex, does not accumulate inequalities between persons when it is systematically applied over the lifespan. So there is no in-principle objection to age-based entitlements, and there is not necessarily any inconsistency with decrying discrimination by age in other settings.

Second, some age-based entitlement programs do not adequately meet the needs of the worst-off elderly. It is not obvious, however, that the way to correct this problem lies in dismantling a system of savings that involves age-entitlements. In general, these problem cases are ones in which more basic problems of distributive justice remain uncorrected despite the presence of what would otherwise be a prudent, age-based savings scheme. In a sense, the *frame* guaranteeing fair income or health-care entitlements is not working, and the otherwise prudent savings scheme cannot be expected to correct these more basic injustices. The risk in eliminating age-entitlements is that we risk dismantling a mechanism intended to promote fair distribution between age groups, and it is the "all-class" aspects of this mechanism that has protected the worst-off elderly from the fate suffered by the poor young. The poor young have had the redistributive programs that are aimed at their welfare drastically reduced in recent years. The poor elderly have been sheltered to some extent from these forces by the all-class nature of age-entitlement programs. Better "targeting" of welfare benefits is always desirable, but that might not be the effect of reforms that dismantle our current, however flawed, solutions to the age-group problem. We may end up only with better targeting of measures *against* the poor. I return to this issue in Chapter 7.

6

The Challenge
of Long-term Care

Issues of Justice in Long-term Care

Long-term care is the wide array of services required by people who have some functional incapacity that limits their ability to care for themselves over a significant period of time, in many cases, permanently. These services, which may be continuously or intermittently needed, include: medical and mental health care, nursing care, rehabilitative therapy, personal-care services, and social services. Long-term care may be institutionally or home-based. Long-term-care beneficiaries can be young or old. As many below age 65 need such services as over age 65, but the likelihood of needing such care increases sharply with age, especially after age 75. Some of these services are provided through public funds or social insurance schemes, but about 65–80 percent of all long-term care, depending on estimates, is provided by family members.

Because the likelihood of needing long-term care increases with age, the aging of society raises urgent questions about our long-term-care system. Some experts suggest that long-term care "may well be the major health and social issue of the next four decades, polarizing society over the next 20 to 40 years" (Vogel and Palmer 1982:v). By 2040, there is likely to be a fivefold increase in the number of people aged 85 and over, and similar increases in the numbers of the very old who are nursing home residents or functionally dependent in the community (Soldo and Manton 1985:286). These changes mean an increase from 2.6 to 13.3 million elderly aged 85 and over, of whom slightly more than 4 million will require personal-care assistance in the community (Manton and Soldo 1985). Society will have to increase its production of long-term-care services from a current estimated base of 6.9 million daily units of long-term-care services to 19.8 million daily units in 2040 (Manton and Lui 1984a, cited in Soldo and Manton 1985). In 2040, persons aged 85 and over will account for one-fifth of those over 65, but they will need over one-half of all long-term care services (Rice and Feldman 1983, cited in Soldo and Manton 1985).

The problem, however, will not be that of merely expanding a basically adequate system; our current system is by no means adequate to handle current needs. If we lack the moral clarity and political will to meet current needs, then the aging of society will indeed push us into crisis. Unfortunately, as many critics of the health-care system in the United States have indicated, long-term care may be its neglected stepchild. Criticisms of the American long-term care system focus on the following central faults:

- It is very difficult for poor patients who need high levels of care, especially Medicaid recipients, to find nursing-home placements;
- The cost of nursing-home care and the eligibility requirements for Medicaid reimbursements drive spouses into poverty, often reducing their ability to maintain independence;
- There is premature institutionalization of many individuals, who could be sustained in less restrictive settings if services were available;
- There are many unmet needs for personal-care and social-support services for frail elderly people trying to maintain independent living arrangements;
- Care in institutions for the elderly is often not aimed at rehabilitation even though some rehabilitation would be possible;
- Families providing long-term care are given few services aimed at relieving their burden, such as day-care centers, supplementary home services, or adequate tax relief or income support to purchase such services;
- There is no adequate market for private or group insurance against the risk of needing long-term care.

A number of factors explain, but do not justify, this neglect of long-term care. Many long-term-care services are not medical. They are the kinds of services a wealthy person might buy to improve quality of life, for example, assistance with shopping, cleaning, or cooking. Such services do not ordinarily carry with them the sense of moral importance or urgency that medical services do. Indeed, in general it is only when long-term care can be "medicalized," because it is serious enough to warrant institutionalization and medical supervision, that social obligations are felt and public resources are deployed.

The public provision of such personal and social services also seems open to special abuse. Eligibility for either public provision of these services or private reimbursement through insurance would have to depend on an acceptable measure of functionality. But such measures are difficult to construct and hard to use in the absence of careful assessment by interdisciplinary teams of health-care and social workers. There is also a tendency to see the disabilities of the elderly not as the result of specific diseases but as the normal course of aging. This view, which geriatricians have fought hard to discount, explains weak efforts at rehabilitation and supplements the prejudice against ascribing moral urgency to nonmedical services. There is also little glamour or prestige attached to the provision of long-term-care services.

There are also important obstacles to the provision of private, individual and group insurance for these services: the difficulty of assessing functionality, the problem of adverse selection that would face any short-term schemes, and the problem of protecting benefits against inflation in any long-term scheme adopted early in life (cf. Bishop 1981 and p. 50 in this volume). In general, family networks have substituted for insurance. Since families provide so much of the necessary long-term care, there has been great reluctance to provide public substitutes that would shift the costs from families to the government. This reluctance to shift costs is supported by a concern that shifting responsibility would undermine traditional moral values and family structures.

Two issues are central to these explanations of the problems in our long-term-care system. First, there is confusion about the moral importance of long-term-care services, that is, about the relative importance of long-term-care services and medical services. Second, there is controversy about how to mesh public obligations to provide long-term care with the belief that families are responsible for caring for their elders. In what follows I shall appeal to the Prudential Lifespan Account of just health care, which provides a unified view of both these issues.

The Moral Importance of Long-term Care

The moral importance of personal *medical* services derives not from their glamour or prestige but from their purpose and function, which is to maintain or restore or compensate for the losses of normal functioning.[1] In general, keeping people normal in these ways is of importance because it affects an individual's share of the normal opportunity range. Our Prudential Lifespan Account refines this general point: We must assess the moral importance of medical services by their impact on the age-relative normal opportunity range (cf. p. 73ff.).

The moral importance of *long-term-care* services also derives from their general purpose and function, which is *identical* to that of medical services. Long-term-care services also maintain, restore, or compensate for the losses of normal functioning. Their importance for purposes of distributive justice is also measured by their impact on age-relative normal opportunity range. It does not matter that the services necessary for adequate long-term care are often nonmedical. Personal-care and social-support services, and, of course, rehabilitative services, whether delivered in the home, the community, or the nursing home, may be exactly what is needed to compensate for impaired function and disability. The disabilities that require long-term care are generally not life-threatening, and people live for many years with them. But they have a dramatic impact on an individual's opportunity to carry out otherwise reasonable parts of a plan of life, including continued independence in living arrangements. Since these disabilities affect such a substantial portion of the elderly population in the later stages of life, it is imprudent to design a system which ignores them and meets only the acute-care crises of the elderly. That long-term care should be so seriously neglected in our health-care system is morally indefensible.

Many advocates of increased home-care services, including personal-care and social-support services, have emphasized the importance of independent living. They have sometimes cited a principle calling for care in the "least restrictive environment" (cf. Callahan and Wallack 1981; Vogel and Palmer 1982). Similarly, others have talked about the loss of dignity that accompanies premature or inappropriate institutionalization. The underlying issue, however, is the loss of opportunity

---

1. More exactly, to maintain or restore species-typical functional organization and, thereby, functioning. Cf. Daniels (1985).

range, which has a direct effect on autonomy and dignity, as well as self-respect. The Prudential Lifespan Account thus provides a sound basis for the view that institutionalization should be a last resort.

The Prudential Lifespan Account also makes it clear why possibly lower costs is not the only—or the most important—basis for advocating alternatives to institutionalization. The literature on the relative costs of nursing-home care versus home- or community-based care is controversial, partly because of methodological problems. There is considerable evidence that some alternatives to institutional care are cost-effective, but this may not be true for some conditions requiring complex care. Nevertheless, protecting independence, and thus crucial elements of an individual's range of opportunities, has an importance that overrides at least some considerations of cost. Of course, cost must enter the picture, since we want a system that is effective in protecting opportunity in the broad range of cases.

Two qualifications to this account of the moral importance of long-term care are in order. First, I assume that at least most of the impaired functioning that requires long-term care for the elderly is the result of specific diseases, which, though more frequent among the elderly, are not normal features of the design of our species. I side here with the broad movement among geriatricians that sees such conditions as diseases and not merely as features of normal aging. I note, however, that this stance ignores a controversy in the philosophy of biology, namely, whether aging is a disease or is part of the design of the species (cf. Caplan 1981). (This qualification is important because my assumption that these conditions are diseases and not part of our design as a species plays a role in characterizing the normal opportunity range for the later stages of life.) Second, there is the problem of extreme cases. Where disability is so severe that services do nothing to compensate for losses of normal functioning, for example, in very advanced stages of Alzheimer's disease, we cannot explain the importance of these services by their effect on opportunity. This problem is not special to long-term care, for medical services face the same issue wherever there is terminal illness. As I have suggested elsewhere (Daniels 1985), in these contexts, other moral considerations, such as beneficence, may require humane care where principles of distributive justice no longer inform us about the relative importance of treatment.

The Prudential Lifespan Account shows that there are social obligations to provide an appropriate array of long-term-care services, and makes it clear that these obligations are of comparable weight to those which govern medical services. We need to say more about the content of these obligations. To do so, we should adopt the perspective of prudent deliberators who are considering the design of our health-care system, while under the constraints described in Chapters 3 and 4.

Consider first the problem of market failure for long-term-care insurance. In Chapter 3 and earlier in this chapter, I noted that there is no significant market for long-term-care insurance, despite the fact that the uncertainty facing the onset and costs of disability makes it an obvious candidate for insurance schemes. Specifically, it would be prudent for individuals to buy contingency claims on the joint risks of disability and other facts, such as the absence of family support or the unsuitability of living arrangements. Yet the potential demand for such insurance has not produced a market for it. Prudent deliberators would meet the obligation to assure access to appropriate long-term-care services by constructing a public or mixed public and private insurance scheme.

Such a social insurance program would constitute the decision to *save* health-care resources at an appropriate rate, given the age profile and the disability/age profile for the society. Specifically, since the aging of our society means that more people will live to experience partial disabilities in the late stage of life, then health-care resources must be saved at a greater rate overall. Moreover, the savings must be in the form of contingent claims on the services appropriate to meeting long-term-care needs. We need to note that a number of public policy proposals have recently been made which constitute steps in this direction. They range from attempts to expand Medicare to cover long-term care to attempts to demonstrate the viability of private group insurance for such services.

Expanding long-term care as an addition to the existing system obviously means expanding resources invested in health care overall, and it means saving such resources at a greater rate. But if resource constraints limit the degree to which we can expand the health-care system, then some hard choices will have to be made. As I have argued, long-term-care services are morally important for the same reasons medical services are. Consequently, choices will have to be made that may reduce expenditures for medical services, even those that marginally extend life for the terminally ill, in order to provide for long-term-care services that significantly enhance opportunity in the later stages of life. The thrust of my discussion of the moral importance of long-term care is that prudent deliberators might well reduce expenditures on medical resources which have little impact on protecting normal opportunity range (for example, lavish rescue attempts to extend the lives of the terminally ill) in favor of significantly improving opportunity for those who need long-term care. It is possible to understand some of the rationing choices made in the British National Health Service, which provides considerable home care to the partially disabled, as a choice made in this spirit. Of course, trading medical services for long-term care is not always a choice between benefits to two distinct classes of individuals. An individual denied lavish rescue efforts on his deathbed may already have benefited from long-term-care services made possible through such rationing. Prudent deliberators try to make this choice from the perspective of what is most important within a life.

One central issue facing us in our role as prudent deliberators is the appropriate division of responsibility for long-term care between private (for example, family) and public measures. Controversy about this issue is part of the explanation for many inadequacies in our current system. In Chapter 2, I argued that it would be morally wrong to protect the supply of family care by legally imposing some set of filial obligations for which we can provide no adequate moral justification. I shall return to the question of filial obligations again in the next section. Here I want to show how prudent deliberators might address the problem of family care, simply knowing that most adult children in their society are willing to provide some long-term care to frail parents, either out of love or a feeling of obligation. The important point that emerges is that not all choices we have to make as prudent planners are part of a *zero-sum game*. What is given to one stage of life, one age group, is not necessarily a loss to another.

Prudent deliberators would want to protect themselves against the contingency that they have no family supports to provide long-term care. They would design the system so that social obligations were met through the appropriate public or private insurance scheme. Given limits on resources, they would want to protect their normal opportunity range by providing the least restrictive forms of long-term care possible. But the more common condition is that some family supports would also be available. Then it would be prudent to arrange institutions so that such support would not be unduly burdensome to family members.

Reducing these burdens is important for several reasons. It means that the care is less likely to be discontinued by families. What care is given by families may then be of higher quality (Dunlop 1980). To the extent that relief services improve the duration and quality of the care provided by families, the elderly are likely to maintain their independence from nursing homes longer. Also the care will be delivered with less stress on family relationships (Doty 1986).

In designing a system through which they will live at each stage of life, prudent deliberators must consider this issue not only from the perspective of the recipients of such care, but also from the perspective of its providers. As family members providing such care, they would be prudent to design a system that gives relief from the daily routine; day-care facilities for the elderly, some social-support services to relieve the family, and, to a lesser extent, tax incentives to reduce the stress on family obligations to younger children—all would make it easier to provide care for elders out of love or filial duty. When family members who actually provide care are asked what sorts of assistance they most require, they put at-home medical and personal-care assistance high on the list along with community day care (cf. Doty 1986). Relief services can also have a significant impact on opportunities available to the primary-care giver. This point may be of special importance in a period when more women, who have traditionally been the primary-care givers for elderly parents, are joining the work force and pursuing careers until late in their own lives. Thus, from the perspective of prudent deliberators, such relief services benefit us at each stage of life.

There are many other issues about the division of responsibility between families and the public sphere that are not addressed by this argument. But the prudential lifespan perspective has led to conclusions that address many of the criticisms of our existing health-care system as noted on page 104. The bias of our system in favor of medical services, and in favor of "medicalizing" long-term care through premature institutionalization, is not defensible on grounds of justice. There are social obligations to guarantee access to adequate long-term-care services, whether institutional, community, or home-based. This realization will require the government to guarantee a public or mixed public and private insurance market for these services. Home and community-based services that preserve independence for the partially disabled (regardless of age) and include rehabilitation to the extent feasible are requirements of the prudent—and just—design of a health-care system. To the extent that society wants to preserve family provision of many long-term-care services, it is prudent, from the perspective of provider and recipient alike, to relieve the burden on families. This point, however, leaves many other issues about the division of public and private responsibility unaddressed. In the next section I shall return to the issue of filial obligations and the division of responsibility.

Filial Obligations and Social Obligations

In Chapter 2 I argued that "traditional" filial obligations should not be used as the basis for inferring our social obligations to care for the elderly. I criticized the Traditionalist position that we could locate in our shared heritage a well-defined set of filial obligations and extrapolate from the burdens they imposed "then" to an understanding of the limits of these obligations now. To a considerable extent, the belief that there is such a tradition is based on a myth. The myth, I argued, masks the actual diversity that exists in our heterogeneous society. Moreover, the burdens filial obligations could have imposed in earlier times are a poor basis for extrapolating to the current situation. Demography and family patterns have dramatically changed. Finally, without suggesting that there is no reasonable basis for filial obligations, I urged that most standard philosophical models for talking about obligation fail to provide obvious grounds for filial obligations. These models thus fail to clarify the limits of such obligations, especially with regard to the provision of long-term care.

I want to return briefly to the issue of filial obligations, for it is important to explain why many people believe they have a real duty to provide care for their elderly parents. In a study of New York caregivers, "family responsibility" was the most frequently mentioned reason for providing long-term care to an elderly relative; it was cited spontaneously by 58 percent of care-givers. The second most common motivation mentioned was love (51 percent), followed by "reciprocity" (17 percent) (Horowitz and Dobrof 1982, cited in Doty 1986). I shall show here how the account I offer of these social obligations, from the perspective of the Prudential Lifespan Account, meshes with one way of providing a basis for filial obligations.[2] What I say here does

2. I have benefited in what follows from a conversation with Christina Hoff Sommers and from Sommers (1986). My account differs somewhat from hers in its notion of background justice and in my suggestion of a contractarian argument. The "conventional expectations" that give content to filial obligations on her account would probably also fall short of obligations to provide extremely burdensome forms of long-term care.

not, however, run counter to my earlier argument that we should not "privatize" our social obligations regarding the elderly.

Families are a fundamental cooperative arrangement within our society. They provide a set of goods of mutual benefit to various family members, at least when families are functioning nonpathologically. The benefits to children are obvious: nurturing and security, education, the development of personality, character and values through normal bonding to parents and siblings and through the practice of intimate, cooperative interaction with others. There are also obvious benefits to parents: Many find childrearing an enriching—perhaps the most enriching—project within their lives. For many, it gives meaning to life and an opportunity to do significant good for others at a level where an individual can exercise significant control. (Neither of these lists of benefits is exhaustive, and it would be easy to add to the list some obvious costs and risks.)

That family structures in our society produce important goods through a form of cooperation does not mean there are not comparable or superior alternatives. In other cultures there may be equally successful or even superior family or collective (community) arrangements for producing these goods. What is important for my argument is that (1) families as we know them work reasonably well, when functioning normally, to produce these important goods; and (2) there is no alternative that is obviously superior and to which it would be relatively costless to switch. I believe both of these conditions are met in our society.

Where a cooperative arrangement produces mutual benefits in the normal case, *legitimate expectations* arise about the importance of reciprocity and the importance of sustaining the kinds of interpersonal relationships which generate these benefits. Certain virtues of character will be important to develop and sustain if the intimate cooperation needed to generate these goods is to be possible over extended periods of time. Moreover, these features of character must allow a flexible response to changing situations as families mature and individuals in them pass through different stages of life. Central among these virtues will be a willingness to act generously to meet the needs of others. Families that are to function effectively in generating mutual benefits will have to encourage the exercise of these virtues and attitudes and members will have legitimate expectations that aid to others will be forthcoming.

These expectations are legitimate even if children did not enter into the family as consenting equals, as actual contractors. The point is that some social structure, such as the family, is a necessary condition for generating a set of goods of fundamental social importance. Any structure, whether family- or community-based, that could succeed in generating these goods would require similar levels of cooperation among children and adults. In turn, such cooperation would require the inculcation of such attitudes and virtues as the willingness to provide mutual aid and to demonstrate respect for others involved in these cooperative structures. The expectation that such aid and respect would be shown is *legitimate* because it would be unreasonable not to accept the necessity of some such set of virtues. This might be recast as a contractarian argument. It is reasonable to accept obligations that appropriate hypothetical contractors would be compelled to impose in designing a just society. Hypothetical contractors would have to select either families or reasonable alternatives in order to assure that essential social goods are achieved, and all these alternatives would carry with them some forms of legitimate expectations, and thus obligations, to provide generous mutual aid. Indeed, if there is no reasonable alternative to existing family structures available to us, hypothetical contractors would be compelled to incorporate structures which generate *some* filial obligations.

Notice that this argument does *not* rest on claiming that all long-standing or typical expectations that arise in traditional structures or institutions are *legitimate*. Institutions that are built on injustice, for example, will not give rise to legitimate expections. The expectations of the racist in the American South that Jim Crow laws should be obeyed were not legitimate ones. This would be true even if there were some mutual benefits that arose between slaves and masters in slave institutions. The expectations for mutual aid that arise within families are legitimate because these institutions are not unjust—even if some forms may be better than others—and because some such social structure, giving rise to similar expectations, is necessary in any case.

The filial obligations that can be based on this sort of view are of course limited in important ways. First, they arise in the normal case. Although they do not require parents to have been "great parents," they do not arise in pathological cases. Second, the obligation to aid others with whom one has intimately cooperated to produce mutual benefits still cannot oblige one to provide *whatever level of aid is needed at whatever cost.* The obligation to aid family members may well require more than obligations to aid one's fellow citizens, but there is nothing about its foundations that means it obliges children to seriously threaten their well-being (or their children's) in aiding parents.

In Chapter 2 I suggested that if filial obligations arose only as a special case of the duty of beneficence, because of the unique position of family members to assist each other, they could still not require the kinds of major sacrifice involved in the extended provision of long-term care. Even though I am now saying that filial obligations arise out of the obligation not to frustrate legitimate expectations of those with whom one intimately cooperates to produce mutual benefits of fundamental importance, it still does not follow that extensive sacrifice by children for their parents will be obligatory. Indeed, it is not clear that there can be *legitimate* expectations of such extensive sacrifice, even if there is legitimate expectation of a great willingness to provide mutual aid.

How do the limited filial obligations I have just described mesh with the social obligations to provide long-term care which derive from the Prudential Lifespan Account? First, the social obligations have a certain primacy. The needs for long-term care must be met even if individuals neglect their family obligations. Moreover, limited filial obligations cannot be expected to meet extensive and burdensome needs for long-term care. Social obligations to meet these needs have to be satisfied.

Second, different societies with different family institutions may inculcate more- or less-extensive family responsibilities. Such societies may then want to divide the responsibility for long-term care differently, and yet each could be just. In societies with strong traditional commitments to caring for the elderly, the fact that there are social obligations to meet long-term care needs does not mean that public measures must rigidly be substituted for family care. At an institutional level, justice leaves considerable room for respecting cultural differences. By the same token, in a heterogeneous society such as ours, no one set of filial obligations should be required by law, as I argued in Chapter 2. At the same time, social obligations can be met in a way that promotes or facilitates the provision of family-based care wherever families feel the responsibility or desire to provide it.

The situation in the United States is serious. The failure to meet social obligations, combined with the desire of some to "privatize" long-term care and resurrect "traditional" family values, actually makes it *more* difficult for individuals to meet family responsibilities.

Some who would provide care, or would do so for longer periods, are unable to meet the responsibilities they feel because appropriate support services are not available. This leads to frustration, guilt, and even rationalization that can undermine an individual's conviction that he or she has such family responsibilities. The frustration and guilt may be felt by frail elderly parents and adult children alike. Even when responsibilities are met, the absence of access to appropriate long-term-care services means that the very families whose integrity and values we are trying to encourage are stressed with burdens of care (cf. Doty 1986), which it is unjust to impose. As society ages, what used to be a relatively rare problem—being part of a family trying to meet a frail, aged parent's needs—will become a very common problem. Meeting social obligations regarding long-term care *so that we may better meet family responsibilities* indeed looms as a most urgent problem during the next several decades.

The New Shape of a Life

The Prudential Lifespan Account, with its emphasis on protecting fair equality of opportunity, views adequate long-term care as a vehicle for extending the "young-old" years as much as possible. A prudently designed system will give partially disabled individuals greater opportunity to pursue their life plans in the later stages of life. A prudent division of responsibility between society and families will lead to better care with lighter burdens at each stage of life. But the aging of society brings with it a broader challenge than merely meeting expanded needs for long-term care.

Increased life-expectancy, especially for the elderly, combined with the demographic shift to an older society, changes the typical *shape* of a life. We can no longer look at the problems of the old as exceptional problems. Each of us individually faces the prospect of planning for life through its expanded late stages. But collectively, society must adjust its institutions and policies to accommodate this new shape. Here quantity changes to quality: When enough lives have the new shape, there is a critical mass of people who have a common interest in making the later stages of life meaningful and productive. The social task—and this involves issues of distributive justice—is to provide an adequate framework of opportunities and means through which diverse individuals can pursue their own views of the good life. The prudential lifespan perspective can help us understand some of the distributive issues that are involved.

One issue that has been frequently discussed by gerontologists is the standard expectation that the elderly, even the energetic and active majority of the elderly, will face the later stages of life without meaningful work. Though anti-age-discrimination legislation has eliminated compulsory retirement at age 65 in many contexts, incentives remain for early retirement, and compulsory retirement still faces many able and productive individuals at age 70. Given the generally good health of the elderly population under age 75, our current policies seem archaic. For many individuals, including elderly individuals, being productive or pursuing meaningful work, is a central element of well-being. Work is much more important than as a mere means to income. Protecting fair equality of opportunity in the later stages of life will require that we develop flexible policies regarding employment of the elderly, including full- and part-time opportunities, as well as community programs using volunteers. This argument for the importance of such policies goes well beyond reducing the economic dependency of the elderly on transfers from the young. It also goes beyond the importance of reducing the rate at which society must save for the later stages of life, though this is an important issue.

Society must further face the educational implications of the fact that life has a new shape. People must be educated to think prospectively about what an additional fifteen or twenty years beyond age 65 should include. We will have to undertake extensive adult education so that people can actively prepare for the projects and plans an extended life makes possible. The new shape of a life makes it more important to provide people with the means and opportunity to revise their plans of life relatively late in life. We must also explore new housing and community living arrangements that harness the cooperative energy of the expanded numbers of the elderly. Many of our attitudes toward education, work, family, and living arrangements are tied quite directly to the old shape of a life. It will be imprudent not to revise these attitudes and the institutions that arise from them. Not revising them will lead to unjust, that is, imprudent distributions of resources between the old and the young.

7

Income Support

Age Groups, Birth Cohorts, and Income Support

Early in this book I distinguished between two problems of distributive justice, which are sometimes lumped together as problems of "generational equity," namely, the problems of justice between age groups and justice between birth cohorts. The problem of justice between age groups concerns the distribution of important social goods among groups of people who differ in that they are at different stages of life. It is a perennial problem that arises regardless of which birth cohorts happen to constitute the different age groups. To solve this problem, I argued, we must consider principles which govern the design of institutions that distribute goods over the lifespan. In developing the Prudential Lifespan Account, and in applying it to the distribution of health care between age groups, I have ignored the fact that different birth cohorts may be affected differently by institutions intended to solve the age-group problem.

This *idealized* account of the age-group problem must now be reconciled with the problem of justice between cohorts. Birth cohorts, after all, differ in size and life history. They are educated, mature, work, raise children, and retire under different economic and social conditions. Such differences between cohorts in general will mean that an ideal scheme of transfers between age groups will require modification or adjustment if cohorts are to be treated equitably. Institutions, such as health-care and income-support systems, must embody solutions to both problems of distributive justice. Of course, we will have to consider what justice or equity between cohorts involves, and not take the notion for granted. In this chapter I shall concentrate on the distribution of income over the lifespan, in part because concerns about equity between birth cohorts have long been a source of criticism of our Social Security system. But the need to reconcile age-group and birth-cohort issues arises in the health-care system as well.

Before we take up the birth-cohort problem and its relation to the age-group problem, it will help to recall how the Prudential Lifespan Account abstracts from the birth-cohort problem through idealizing assumptions. I have imagined that we are concerned with institutions which operate over the lifespan. We are to think about how these institutions distribute important goods *within lives rather than between persons*. The problem of distribution between the young and the old can be viewed as a purely prudential problem only if this idealization is made. Purely prudential reasoning cannot solve problems that involve distributions across the boundaries between people. To exclude interpersonal transfers, I have had to *frame* the age-group problem. Specifically, I assume that *fair shares* of health care or income are being distributed over the lifespan. Similarly, I assume that prudential agents reasoning about these distributions must imagine that they will have to live through each stage of life under the arrangements they select. This part of the frame in effect rules out knowledge of how old one already is. It means that choices will not be biased in favor of one birth cohort over another.

These idealizations which rule out interpersonal transfers are necessary to yield a solution to the age-group problem. But they do not reflect the way in which institutions actually affect different cohorts which pass through them. In the past, for example, many have complained that Social Security *benefit ratios*, the ratio of benefits received to contributions made, varied dramatically between birth cohorts. *Early entrants*, who received benefits without having paid payroll taxes over much of their working life, had windfall returns compared to later entrants in the mature Social Security system. Similarly, many predict that benefit ratios will be much worse for the baby boom cohort than for current retirees even in the mature system. Our idealized solution to the age-group problem thus ignores real interpersonal transfers, and such transfers raise questions of distributive justice not addressed by our model of prudential deliberation. One form the birth-cohort problem takes, then, is the question: What inequalities in average benefit ratios between cohorts are inequitable?

The idealizations in our prudential lifespan model need to be qualified in another way, for they ignore a second type of interpersonal transfer which takes place within the Social Security system. The Social Security system yields different benefit ratios for different economic subgroups because it carries out a *redistributive* and not merely a savings function. This redistributive function means we cannot assess the Social Security system by reference to only one social goal or one principle of distributive justice. In effect, the interpersonal transfers required by more general principles of justice work *within* the institution whose main function is to solve the age-group problem. The boundary between the frame and the age-group problem is thus blurred within the main institution concerned with income transfers over the lifespan. When we ask whether actual institutions function in accord with ideal principles, we must take care to examine all the functions of the institution.

An important focus of this chapter will be the growing appeal in public policy discussions to the concept of "generational equity." This concept underlies complaints about inequalities in benefit ratios between birth cohorts. In the extreme, of course, these concerns rest on predictions about insolvency of the combined income support and Medicare systems as a result of the aging of our population. The failure of such systems would surely be unfair to cohorts which contribute but never receive benefits. But it will not take outright collapse of these systems to produce significant inequalities in benefit ratios. The Prudential Lifespan Account can clarify what is valuable and what is misleading about these appeals to "generational equity" in public policy analysis.

The Prudential Lifespan Account and Income Support

Before discussing the birth-cohort problem, I want to comment briefly on the implications of the Prudential Lifespan Account for the distribution of income between age groups. How would it be prudent, from the perspective of our hypothetical rational deliberators, to allocate income over the lifespan? The answer to this question should tell us what distribution of income between the young and the old is fair. It should tell us how institutions should be designed to facilitate income transfers between age groups.

Our deliberators operate within the frame I described earlier in the chapter (and in Chapters 3 and 4). They must suppose that fair shares of income are to be allocated over the lifespan and that they must live through each stage of the scheme they devise. I have also argued that these prudent deliberators should not be the fully informed agents economists standardly employ in their models. Rather, these deliberators do not know the details of their plan of life and must instead reason about their well-being by reference to a Rawlsian index of primary social goods. The argument for this constraint, however, is not Rawlsian; rather, it is justified by an appeal to the classical theory of rational prudence. To demonstrate an equal concern for all parts of their lives, the prudent deliberators should not base their choices on the details of the plan of life they happen to hold at the time of choice. Since we revise plans of life, often in fundamental ways, as we age, we must avoid biasing our choices in favor of preferences we happen to have at a given stage of life. In effect, for the purpose of solving this problem of justice, prudent deliberators are barred from thinking about how to maximize their happiness or the overall satisfaction of their desires. Instead, they maximize their well-being as measured by an index of primary social goods, though we shall here let income stand as proxy for the whole index. (The justification for these constraints is provided in Chapter 3).

Under these constraints, prudent deliberators would have to reason as follows. They cannot expand their lifetime income share by allocating it in certain ways, for example, by setting aside income early in life and investing it heavily in their own human capital or otherwise. Such investment strategies are already accommodated within the notion of a (lifetime) fair income share, or so I am supposing. Similarly, they cannot argue in favor of allowing inequalities in income levels between stages of their life provided that such inequalities work to make them maximally well-off (as measured by income) during the stages of life in which they are worst-off. It is important to see why this familiar, Rawlsian argument will not work here (cf. Rawls 1971).

This form of reasoning is used by the hypothetical contractors in Rawls's Original Position. They choose Rawls's Difference Principle to govern income inequalities *between* persons. Cooperation among people with unequal shares may generate a larger social product, for example, through certain incentives. This larger product, though divided unequally, can make the worst-off individuals better off (as measured by the index) than they would have been had they insisted on equal shares. But this reasoning is blocked by the assumption that lifetime shares of income are already fixed as fair shares. Inequalities in income levels between stages of life work only as a zero-sum game, making one period of life better off at the expense of another period or stage in life.

If we knew that a plan of life for an individual contained preferences which at all stages of life would make the unequal distribution preferable because it would increase lifetime satisfaction, then that would be a better distribution for that individual. But we cannot know that plans of life will cohere in this way, and in general they do not. It is this fact that led us earlier to block out information about plans of life in order not to bias planning in favor of what the young think is prudent. So the prudent course of action would be to allocate resources in such a way that income (standing proxy for the complete index of primary goods) would remain roughly equal over the lifespan. (I say "roughly equal" because we will want to adjust income in a way to be explained.) The Income (or Standard of Living) Preservation Principle, as I shall call it, ensures that institutions facilitate income transfers over the lifespan in such a way that the individual has available to himself, at each stage of life, an adequate income to pursue whatever plan of life he may have at that stage of life. Of course, "adequate" is here relative to his lifetime fair income share.

To apply the Income Preservation Principle to our income-support system, we need an important qualification. Income must be adjusted if it is to count as a measure of well-being at different points in the lifespan, for we are really interested in *standard of living*. Specifically, we must adjust income to accommodate some general facts about how income *must* be spent by individuals, that is, facts about the needs for income at different stages of life *given institutional arrangements in the United States*. For example, if income spent on raising children, including advancing their educational opportunities, is at least in part inelastic and represents a "durable good" that produces benefits to parents over the lifespan, then we should adjust pre-retirement income levels by subtracting some portion of income spent on children. Similarly, if out-of-pocket individual expenditures on health care rise substantially with age, as they typically do in the United States, then we must adjust the income of the elderly to account both for Medicare payments and out-of-pocket costs. We are interested in the relationship between income and standard of living because equal income levels at different stages of life may not represent equal standards of living.

A similar retreat from unadjusted measurements of income is made by economists when they want to measure the effects of our income-support system on well-being over the lifespan. When economists discuss the adequacy of income-support levels under Social Security, they invoke the notion of a *replacement ratio,* which is a ratio of *some* measure of retirement income to *some* (perhaps different) measure of pre-retirement income (see Boskin and Shoven (1984:2). Difficult methodological issues, which are not always morally or ideologically neutral, surround the decision to use a particular measure of income in calculating replacement ratios. For example, these ratios are very sensitive to whether we use peak-earning-year income levels or average-earning-year income levels. Similarly, they are very sensitive to whether or not we adjust income levels to exclude certain childrearing expenses or health-care expenses, on the grounds that these uses of income are inelastic responses to needs which should not be pooled with other straightforward "consumer" uses of income.[1]

The Income Preservation Principle appears to favor a replacement ratio of 1 between pre- and post-retirement income. To determine whether our income-support system is in compliance with this principle, we would have to resolve these empirical disputes, which is not my task here. Moreover, the contribution of Social Security benefits to overall post-retirement income varies with income level. Social Security benefits comprise a decreasing share of post-retirement income as income levels increase. Thus the application of the Income Preservation Principle to the design of the Social Security system is further complicated by its differential contribution to income support for different economic subgroups. Since Social Security is embedded within a more comprehensive public and private system of income transfers between age

1. Cf. Boskin and Shoven (June 1984) for an excellent discussion of the methodological issues and some calculations of the effects on replacement ratios of using different income measures.

groups, including a full array of individual and group pensions and other sources of income, we cannot apply the principle to Social Security in isolation.

The Prudential Lifespan Account in general suggests that post-retirement income levels (or standard of living) should approximate pre-retirement income levels (or standard of living), across the board for all levels of fair income shares. Is this principle satisfied in the United States? We can answer this question only in part. There is some evidence that our income-support system, lumping together Social Security and other sources of pension income, preserves standard of living on the average. Aaron (1982:70) summarizes evidence which suggests "that the elderly on the average are able to sustain consumption during retirement about equal to the average consumption they could achieve over their lifetime." There are, however, important subgroups of the elderly, for example, widows or the very old, whose income levels fall well below their lifetime averages.[2] In any case, the principle our prudent deliberators would choose applies only to *fair income shares*. We have not addressed the question whether income distribution in the United States is in general just. There are good reasons to think it is not, but I cannot address this complex moral and empirical question here without abandoning my subject. Thus we cannot say whether the Income Preservation Principle generally holds true in our system.

One point about the income redistributive effects of Social Security is relevant, however. Different economic subgroups within each birth cohort fare differently in terms of benefit ratios. Lower income groups enjoy higher benefit ratios than upper income groups, which gives the appearance of an income transfer between subgroups within a cohort. Of course, the real transfer is between high-income groups in later cohorts and low-income groups in earlier ones. Assuming stable institutions operating over the lifespan, this is equivalent to transfers between subgroups within cohorts.

The point that is relevant here is that such income redistribution— which *seems* to go beyond the mere savings function of Social Security—may nevertheless be necessary if the Income Preservation Principle is to be satisfied. That principle governs the whole system of transfers between age groups—all public and private income-support

2. See Boskin and Shoven (May 1986) for a detailed study of pockets of poverty among the elderly.

measures. Some degree of "progressive" taxation and "progressive" benefit structure within Social Security itself may be needed to correct for "regressive" measures elsewhere in the system of transfers between age groups. Thus a Social Security system that provided less income redistribution when considered by itself might, when seen as part of the total income-transfer system, actually increase income inequalities over the lifespan. Thus the redistributive function of the Social Security system can keep the whole income-support system from violating the Income Preservation Principle more widely.

This point is important because some critics of the Social Security system believe it should not be used to carry out multiple functions, that is, to provide for income redistribution as well as savings or annuities. They believe it "sneaks" in redistributive effects under the guise of its savings function. But regardless of whether the public is adequately informed of the redistributive effect, it is important to see that the redistributive function may result only in a more acceptable *savings* outcome. That is, it may only *appear* to be redistributive, while it is actually correcting biases that make it more difficult to satisfy the Income Preservation Principle in the total income-support system. When some critics of the Social Security system (see Boskin, Kotlikoff, Puffert, and Shoven 1986) point to significant differences in replacement ratios and to transfers within cohorts, we should not be tempted to ask: Is this pattern equitable or just? We cannot answer this question about Social Security in isolation from the rest of the income-support system. We must look at the overall effect on the standard of living as a whole to see whether this part of the system acts in ways that conform to the Income Preservation Principle. This point is independent of whether or not our overall income distribution is just or fair.

Justice Between Birth Cohorts

In what follows, let us assume that we have a solution to the problem of justice between age groups. If the Prudential Lifespan Account is correct, the Income Preservation Principle should govern institutions involved in income support and the Age-Relative Fair Equality of Opportunity Principle should govern the design of our health-care system. We still must ask how different birth cohorts fare when they pass through institutions intended to meet the requirements of justice between age groups. What does justice require in the treatment of birth cohorts? What kinds of inequalities in the treatment of birth cohorts are still equitable or fair? In particular we want to know when benefit ratios are equitable (where the *benefit ratio* is the ratio of benefits received to contributions made).

In general, how different birth cohorts will fare in these institutions over their lifespans will depend both on features of the institutional design and on such important social variables as demographic changes and variations in rates of real economic growth. For example, a pay-as-you-go system (like our Social Security system) relies on direct transfers between birth cohorts. It is particularly sensitive to changes in birth rates or the ratio of retirees to contributors. An income-support system that relies more directly on vested savings by each cohort is less sensitive to these changes, but it is more sensitive to others. It may provide less protection against inflation or increases in life expectancy, and it may leave each birth cohort to benefit or suffer from important fluctuations in economic conditions during crucial working years. In what follows, I shall focus attention on the ways in which our pay-as-you-go Social Security and Medicare systems accentuate *inequalities* between birth cohorts in the face of projected demographic changes. This discussion of the kinds of issues involved is meant to be illustrative, not comprehensive.

It will help to compare what I am calling the birth-cohort problem with a problem that resembles it in some ways, the problem of obligations to future generations. Both of these problems involve questions about net transfers of resources from one group of persons—one cohort or generation—to another. In this regard, they both differ from the age-group problem. Transfers from one age group to another can be thought of as allocations within a life rather than between persons. The problem of obligations to future generations can be expressed in this way: Does justice require one generation (birth cohort) to refrain from depleting resources—both the environment (nonrenewable resources) and accumulated capital—so that later generations can have the means to live a decent life? This formulation is sometimes expressed as the question: Is there a *Just Savings Principle* between generations?

Approaches to the problem of a Just Savings Principle are controversial. As Rawls (1971) formulates the problem, he is primarily concerned with *preserving* adequate capital and nonrenewable resources so that successive generations are in a position to maintain institutions of justice. How much must be saved if successive generations are to be able to construct a framework of institutions that provide them with a just distribution of such primary goods as liberty, opportunity, and income? It is to solve this problem that Rawls invokes the device of a thick veil of ignorance. We do not know which generation we will be in when we are choosing our principles of justice. Moreover, he imposes a motivational constraint on parties making the hypothetical contract: They are concerned about the well-being of a generation or two in each direction (from grandparents to grandchildren). Contractors under such constraints, he argues, would prudently grant each generation an equal claim on resources necessary to maintain institutions of justice. In this way, the just savings rate acts as a constraint on other principles of justice, such as the Difference Principle. Thus no society can maximize the well-being of its worst-off members unless it has set aside the resources required by the Just Savings Principle.

Other philosophers have approached the problem differently. Brian Barry (1978:226) suggests that if people in one generation have concerns about the well-being of those in another, such as their children or grandchildren, then because the well-being of such descendents depends on the kind of world they live in, the welfare of the next generation should be thought of as a public good. Coerced saving would be required to preserve that public good. He thinks it problematic, however, that what justice requires toward future generations is contingent on what sentiments people in current generations happen to have. This argument is like Rawls's in making justice dependent on either actual or hypothetical sentiments toward children. Barry (1978:243) also suggests a quite different argument. Justice requires that we preserve equality of opportunity, and this means we must preserve a range of opportunities for future generations comparable to the range we currently enjoy. Barry's argument, however, supposes that the principle of equal opportunity applies across generations. This supposition, however, is part of what is at issue when we ask about obligations to future generations: Are we obliged to protect *their* opportunity? Barry's argument appears to beg the question.

I shall offer no solution to the problem of a Just Savings Principle, not only because I do not have one, but also because I think we can separate the question about equity between birth cohorts from it. Our problem about birth cohorts arises in the context of institutions which transfer income or aid-in-kind between age groups so that their *consumption* will yield just income-support and health-care distributions over the lifespan. The problem of what each cohort must refrain from consuming so that later cohorts can live decent lives arises regardless of whether birth cohorts experience unequal benefit ratios. The birth-cohort problem that concerns us is not a special case of the problem of finding a Just Savings Principle. A Just Savings Principle would not necessarily tell us how to solve the birth-cohort problem, but I believe we can solve the latter without having the former.

Our birth-cohort problem should be approached separately, as follows. Each cohort wants institutions that solve the age-group problem effectively. This is true because each cohort ages and has an interest in solving the age-group problem. But institutions that solve the age-group problem operate under considerable uncertainty. There is uncertainty about population- and economic-growth rates, as well as about technological change, which further affects productivity. Errors are likely to abound and inequalities in benefit ratios between cohorts will arise as a result. But institutions that can solve the age-group problem must remain stable over time. They must weather the political struggle that will result as a result of unjustifiable or unacceptable inequalities in benefit ratios. Such institutions will be able to survive the struggle among coexisting birth cohorts only if each cohort feels it has a stake in preserving them.

Notice that there is an important difference here between our birth-cohort problem and the problem of a Just Savings Principle, at least where the latter is concerned about obligations to distant, future generations. Current generations have a power over distant future ones that contiguous cohorts cannot exercise over each other. Each birth cohort will have an interest in preserving institutions that solve the age-group problem—and such solutions will thus be stable over time—only if there is an underlying commitment to strive for *equity* in benefit ratios.

The practical target for this commitment we can take to be *approximate equality* in benefit ratios. Nevertheless, uncertainty obtains. Unplanned and unpredicted changes in population growth rates or productivity (and real wages) will affect rates of return on the taxes (both individual and employer) that are paid. Economists have shown that in a pay-as-you-go Social Security system, in which no reserves accumulate, there is an implicit rate of return on tax payments that equals the sum of the rate of growth of the labor force and the rate of increase in wages (see Aaron 1982:76; cf. Samuelson 1958 and Aaron 1966). As a work force grows larger and richer it *can* pay taxes sufficient to finance benefits for retirees that are greater than the taxes the retirees paid while working. As population and economic growth rates change, however, there will be changes in the implicit rates of return for different cohorts. Moreover, the question remains, who should benefit from such changes as increased productivity? Should some of the benefits go to retirees, or should they be kept—in the form of lower Social Security tax rates—by the work force? Similarly, if growth rates drop, should the burdens of such decreases also be born by retirees?

The argument has been made that some portion of increased productivity should be shared with earlier cohorts not now in the work force (cf. Spengler and Kreps 1963). Their contributions, in the form of investments in education or research, may have contributed to the increased productivity of later cohorts. Conversely, if productivity declines, because there has been too much consumption and not enough investment in education or research or protection of natural resources, then earlier cohorts ought to share the burden of economic decline.

The framework on which these points rest, however, emphasizes *desert* as a distributive principle. Rewards or entitlements should be proportional to the contributions one has made. This framework has other implications, too. If productivity increases because of the special energy of the current work force, then current retirees would seem to have no claim on increased benefits. If productivity drops, not because earlier cohorts invested foolishly, but because the current work force grows lazy or mismanages foreign affairs, then current retirees do not deserve to share in the losses. Must benefit ratios depend on disentangling these sources of change? It is hard to see how a stable system could incorporate such factors into its scheme of benefits.

We might try to cut through some of the complex issues of desert by appealing, as I did earlier, to the interest each cohort has in providing for stable institutions that solve the age-group problem. Cohorts must cooperate to achieve such stability. But cooperation will require some *sharing of risks* across cohorts. In general, the burdens of economic decline and of living through unfavorable retiree/employee ratios must be shared, as must the benefits of economic growth and favorable retiree/employee ratios. This suggests again that approximate equality in benefit ratios should be a practical target of public policy, if not a hard and fast rule.

We operate an income-support system in a *nonideal* context. It will always encounter various sorts of "interstitial" equity considerations that are generated by both great uncertainty and political expediency (cf. Barry 1965: Chapter 9). A good example is the tremendous windfall in benefit ratios offered the early entrants into the Social Security system in the United States. They paid taxes for very little of their working lives, but received lifetime benefits. Nevertheless, attempting to lower that ratio might have undermined political support for the Social Security system as a whole; that, in turn, would have meant foregoing the introduction of an institution that was arguably necessary to solving the age-group problem. Similarly, in the United States, no reserve fund was ever generated which was significant enough to cushion the effects of recent declines in real wages or projected declines in the ratio of retirees to workers. Politicians were afraid to raise tax rates without pairing the increases with benefit increases. And conservative politicians feared that socialism would follow if large capital funds were controlled by the government (cf. Derthick 1979). (Ironically, it is now conservative politicians who tend most sharply to criticize the unequal benefit ratios later cohorts have.) The point is that some compromises with approximate equality in benefit ratios will have to be made for periods of time in order to establish an institution that on the whole makes the system more just, in this instance by helping to solve the age-group problem.

I noted earlier that issues of equity between birth cohorts arise in the context of health-care as well as income-support institutions. As with income-support schemes, there will be a bias in favor of early entrants. Early entrants into Medicare received benefit ratios far greater than any that will be enjoyed by later cohorts. But there are other forms of inequity between cohorts. Recall our Scheme A (see Chapter 5), in which renal dialysis was rationed by age. An elderly person might complain about Scheme A by saying that it is not really fair to his cohort. His cohort never had the benefit of increasing its chances of reaching a normal lifespan because the technology now being denied it, dialysis, was also not available in its youth.

Two points might be made in response to this complaint. First, it might be argued that each birth cohort is treated equally in the following way. At some point in its life, each cohort will be denied the best available life-extending technologies, but at all other points in its life it will have a better chance of receiving them. To be sure, the particular technology which is denied, in this case dialysis, may not be the very one that the cohort had a better chance of receiving, but there is a fairness in the exchange. Still, if technology improves very rapidly, the bargain is not quite as favorable from a prudential perspective as it might have seemed when we ignored the (possible high) rate of technological change. A second point is more general. Some changes, for example, those in technology, are at least as difficult to predict as the other factors that lead to errors (differences in benefit ratios) in institutions intended to solve the age-group problem. We may be even more prone to error in the health-care setting than in income-support programs. Given the overriding importance of stability in such institutions, however, considerable tolerance for error must obtain.

Intergenerational Equity
and the Social Security System

Institutions that distribute important social goods over the lifespan, such as the Social Security system and Medicare, must be governed by principles that solve both the age-group and birth-cohort problems. To solve the age-group problem, I have initially ignored the birth-cohort problem. But institutions that ideally solve the age-group problem operate under varying demographic and economic conditions and under considerable uncertainty about these variations. These factors lead to inequalities in the way birth cohorts fare. Nevertheless, birth cohorts have an interest in achieving stable solutions to the age-group problem. To achieve such institutional stability there must be a commitment to securing equitable treatment of different cohorts. If one cohort seeks terms too much in its favor, say when it is young, it will very likely pay the price when it is old; similarly, if it seeks too much when it is old, it will risk rebellion from the young. Therefore, I have suggested that approximate equality in replacement ratios should be the practical target of public policy regarding "mature" institutions. Since, however, different cohorts must be willing to share risks, they must also tolerate errors, or departures from approximate equality.

This approach to the birth-cohort problem should help us clarify some criticisms of the Social Security system which have appealed to the concept of "generational equity." Unfortunately, as I noted in Chapter 1, three distinct problems are sometimes confused when appealing to this concept. The concept is sometimes used to address the age-group problem, for example, when critics complain that the elderly are faring well in public budgets at the expense of children. My Prudential Lifespan Account avoids this competitive perspective. Similarly, it avoids the crude notions of equity, such as equal per capita expenditures of public funds on the elderly and the young, which have been associated with the concept of "generational equity" (cf. Chapter 5).

Some critics have appealed to "generational equity" as a way of raising policy questions about obligations between generations, that is, about the problem of a Just Savings Principle. This is a useful focus for public discussion and a legitimate way to invoke the concept of "generational equity." Critics of current policies have in mind the current lack of investment in education and in research and development. These policies can reduce productivity for birth cohorts to come. Similarly, these critics are concerned about the depletion of natural resources and and environmental quality and about the impact these factors will have on the productivity of future generations (cf. Longman 1985). Similarly, our public budgets finance current consumption, including defense expenditures, through deficits which impose debts on future cohorts. All of these concerns are central to the problem of a Just Savings Principle. They are little discussed but very important issues of public policy. I have not addressed these issues while discussing the birth-cohort question, but this hardly plays down their importance. They deserve full discussion, though I cannot consider them in this book.

In this section I shall be concerned with the third problem falling under the umbrella of "generational equity," namely, those issues which coincide with the birth-cohort problem as I have rather narrowly defined it. Critics of pay-as-you-go distributive schemes like Medicare and Social Security have focused attention on the unequal benefit ratios enjoyed by different birth cohorts. We want to consider arguments based on these inequalities in light of my approach to the birth-cohort problem.

One source of complaint about intergenerational inequity focuses on the windfall benefits or returns enjoyed by early entrants into the immature Social Security system. Partly as a result of the immaturity of the system, all cohorts retiring until now have received total benefits worth much more than the combined individual and employer taxes paid on their behalf (assuming discount rates up to 4 percent, cf. Aaron 1982:73, citing Moffitt 1982). In contrast, younger workers may find that the value of benefits will not exceed the value of taxes paid (assuming discount rates of 3 percent). Similarly, the internal rate of return, that is, the discount rate that would equate benefits with the sum of employer and employee taxes paid, has declined from a high of about 20 percent in the 1950s to 8.5 percent in 1977. This rate will decline further (Aaron 1982:73, n.9).

These inequalities between early and late entrants into a pay-as-you-go system are not necessarily unjust or unfair, as I suggested earlier. Inequality does not automatically translate into inequity, even if equality in benefit ratios is a practical goal for a mature system. Special problems are raised by the introduction of new institutions intended to make a system more just. Social Security, which was intended to solve the age-group problem, faced just such problems of introduction. Sometimes we must accept inequalities that would otherwise count as inequities as the cost of reform, for example the price that must be paid if political support for the reform is to be marshalled. Such inequalities should be ignored.

More serious issues are raised by the claim that there will be significant intercohort inequalities within our mature Social Security system. For example, Russell (1982, cited in Aaron 1982:74), calculates a real internal rate of return for cohorts of workers who reach age 65 between 1960 and 2000 that ranges from 2.6 to 4.9 percent. Similarly, Leimer and Petri (1981, cited in Aaron 1982:74) used different assumptions to project real internal rates of return for workers at age twenty-two ranging from 3.7 percent for workers who entered the labor force in 1960 to 2.5 percent for workers entering in 2000. These calculations about internal rates of return must be reconciled with projections made by other researchers who find that young workers now in the work force and future entrants will pay more in taxes than they will receive in Social Security benefits (see Boskin and Shoven 1984, and Aaron 1982:75). These projections all depend on important methodological assumptions, especially an assumed real discount rate of 3 percent (that is, the value of future taxes and benefits is discounted at a real rate of 3 percent). As Aaron points out (1982:75), these projections are compatible with calculations of positive internal rates of return, provided the internal rates of return are lower than the discount rate.

Several distinct kinds of arguments have been based on these figures. First, some argue that because there are significant inequalities in benefit ratios in the mature system, "generational equity" requires that we eliminate them by modifying the way we finance the system and the benefits it pays. This argument leaves open what we must do to eliminate the inequalities. Second, some argue that we should abandon a pay-as-you-go system in favor of one that encourages more savings, since the return on investments in the existing system for future cohorts is less than would result from investments in markets for stocks or real estate. Third, some may argue that individuals, or cohorts, should abandon the pay-as-you-go system since they will not be getting their money's worth out of their contributions. We shall consider these arguments in turn.

The first argument has some merit, though it must be applied judiciously. Stability of the system over time requires that cohorts passing through it have a commitment to equity in the treatment of cohorts: Practically, this involves a target of approximate equality in benefit ratios. Of course, projections are dependent on assumptions about population and economic growth rates, and there is room for error here. Still, where we have reason to expect systematic and significant departures from equality in benefit ratios, we should seek to correct the financing scheme.

One proposal is to build a reserve fund by raising taxes while refraining from distributing the increases as benefits to current retirees. Similarly, one might not distribute to current retirees all increases in revenues that result from the increases in real wages of current workers. The resulting revenue again could be used to build a reserve fund. Such a reserve fund could finance enhanced benefits for cohorts now projected to suffer very low rates of return. It would provide an expendable cushion to soften the effects of swings in the ratio of retirees to workers. This reserve fund need not involve a basic transformation from a pay-as-you-go-system into one that relied on vested assets; it is but a variation on the theme of cash-flow financing. I cannot here discuss the merits or the economics of any specific proposals, since that would take me too far afield. As Aaron also notes (1982:77, n.20), the value of these proposals is to raise the idea that a concern for equity between cohorts should be explicit in discussions of Social Security financing.

The second argument is not really about the implications of inequalities in benefit ratios, though it draws some of its force from the contrast between future and present benefit ratios. In the past, favorable population growth rates yielded reasonably high rates of return. As noted earlier, in a pay-as-you-go system the implicit rate of return workers can enjoy on their taxes is equal to the sum of the rate of increase in real wages and the rate of growth in the labor force. In periods when the rate of return exceeds what might result from investment in real estate or stock markets, the pay-as-you-go system is attractive. But if we can predict implicit returns for present and future cohorts of workers that are below returns that could be made from other kinds of investments of their tax dollars, then it seems these cohorts would be better served by a system that encouraged savings which could be invested in these alternatives. This argument requires the assumption that our pay-as-you-go system has a significant negative impact on savings, a view which is quite controversial.[3] But even if a pay-as-you-go system has a negative effect on savings, there are problems with this argument.

Suppose that additional real investments would yield a rate of return exceeding the projected rates of population and productivity growth, thus exceeding what a pay-as-you-go system could return. It still remains an open question whether increased investments should come from the dismantling or refinancing of Social Security or from other strategies to encourage savings and investment by individuals and the government (Aaron 1982:77). Even if in the long-run, reduced savings inhibit investment, in the short-run many other factors affect the rate of investment. Remodeling or dismantling our pay-as-you-go system is thus not necessarily the appropriate means for increasing investment. As Aaron points out (1982:78), we should not confuse a general theorem of growth theory, that we should invest where we can get the best return, with conclusions about the specific means for enhancing investment.

The third argument, that cohorts facing low rates of return are not getting their money's worth, is also problematic. First, important features of the benefit package provided by Social Security, such as its indexing to protect against inflation, are not replicable in any private market.[4] Social Security solves a standard problem facing annuities: We do not have to decide just how much to save given the uncertainty about how long we will live. Therefore we cannot easily match this feature with alternative investments. Second, real rates of return in alternative

3. See Aaron 1982 for an excellent discussion of the methodological roots of the controversy surrounding claims about the effects on savings of Social Security.

4. Aaron 1982:76 notes that a careful selection of securities can yield a portfolio that behaves as if it were indexed. He points out that it may not continue to do so, and that its real rate of return would be close to zero, which is well under the standardly used real discount rate of 3 percent.

markets are quite variable. A system that relied on such investments, for example, through vested assets of retirees, might make each cohort subject to more significant risks than those involved in the fluctuating ratios of retirees to workers. Third, there are important benefits to be gained from a public transfer system between age groups, such as Social Security, which cannot be realized through private alternatives. For example, there are benefits that result from reducing the dependency of retirees on private family resources. A public system of transfers to the elderly frees adult children to pursue the well-being of their own children and is less demeaning to, and generally preferred by, the elderly. Social Security stabilizes family life by minimizing the risks that would attend private transfers (cf. Kingson, Hirshorn, and Harootyan 1986:24). Thus our current system accomplishes social goals that would not be met by proposals to "privatize" the system.

Not all criticisms of Social Security which emphasize "generational equity" call for dismantling the system in favor of more extensive reliance on private savings and pensions. But it is not an accident that some who favor "privatizing" the system are also attracted to the talk about "generational equity." Libertarians, for example, believe that all coercive welfare redistributions are unjust. On this view, individuals are entitled to the results of free exchanges they make, whereas redistributions take the results of such exchanges without the consent of all parties. Accordingly, libertarians support the concept of a minimal state, one that refrains from interventions in markets, especially interventions which coercively transfer goods from one set of individuals to another. These libertarian views would lead those who hold them to complain about generational inequity. Net transfers from one birth cohort to another, administered through a coercive taxation scheme, constitute a type of nonvoluntary welfare redistribution. Unequal benefit ratios, to the extent that they reflect net transfers of resources between cohorts, are a special case of the injustice or inequity that results from all coerced welfare redistributions. Thus a catchword, like "generational equity," can unite people who actually have quite different concepts of justice or equity in mind. A concern for "generational equity," in the sense of equity between cohorts, does not necessarily imply an interest in "privatizing" Social Security or Medicare.

There are some general lessons to be drawn from these arguments about inequalities in benefit ratios and "generational equity." On my approach, at least in the case of health care and income support, the solution to the age-group problem is basic and the solution to the birth-cohort problem is secondary, though both are important. Solving the birth-cohort problem requires "fine tuning" the institutions which solve the more basic problem. One reason for this priority, which I noted in Chapter 1, is that solving the birth-cohort problem by itself does not answer the fundamental question: Which transfers among stages of life are ones we want social institutions to help us make?

Suppose we were disturbed by inequalities in benefit ratios and we viewed them as inequities intrinsic to pay-as-you-go systems. As a result, we decide to dismantle Social Security or Medicare and seek a system in which each cohort depends on its own vested savings for income support and health care. That is, we "privatize" our institutions, at least between cohorts. We would still have to answer the question: How can social institutions facilitate adequate types and rates of saving? That is, we are back to the age-group problem, but we must now solve it by relying only on the resources of one cohort. Moreover, we are ruling out an important advantage offered by a system that involves intercohort transfers, namely, that it tends to share risks more widely over time. We cannot take advantage of the fact that an equitable form of risk-sharing would be much more desirable than the results of this form of "privatizing." Concerns about intergenerational equity should not drive us away from promising solutions to the age-group problem, since we cannot escape it in any case.

Another lesson of these arguments about intergenerational equity is that they focus too much attention on Social Security in isolation from the rest of the system that accomplishes transfers between age groups. For example, one factor that lowers overall rates of return for certain cohorts is the very low rate of return the highest income groups derive from their Social Security contributions. But we already saw, in discussing income support and the age-group problem, that we must examine the whole transfer system and not just Social Security in isolation. Within the overall system, including private pensions and other sources of income, higher income groups would fare better. Overall, then, the aggregate rate of return to a whole cohort underestimates the importance of Social Security as a solution to the age-group problem for lower economic groups within the cohort.

Examining Social Security or Medicare in isolation, through the lens of "generational equity," may mislead us in other ways. We can be misled into leaving unconsidered many other features and policies of the society as a whole. We can be misled into thinking that we are playing a zero-sum game in which public resources and strategies for solving certain problems are fixed and in which the gains of one cohort are necessarily the losses of another. We can be tempted into ignoring the ''common stake'' that different cohorts have in stable solutions to the age-group problem (cf. Kingson, Hirshorn, and Harootyan 1986).

These errors amount to *taking the birth cohort problem out of context,* specifically out of the context of other issues of distributive justice. To keep it in context, I have deliberately restricted the birth-cohort problem in important ways: It is the problem of achieving equitable treatment of distinct cohorts as they pass through institutions that are intended to solve the age-group problem in an otherwise just society. To solve the age-group problem, I not only idealized away the birth-cohort problem, but I assumed we are concerned with an otherwise just system. But just these assumptions about background justice are ignored when complaints about ''generational equity'' become the primary focus of public policy analysis.

Appeals to ''generational equity'' often ignore this issue of background justice or context. Coupled with complaints about unequal benefit ratios and the fate of the baby boom generation, we also hear the suggestion that Social Security and Medicare give too much to the elderly at the expense of children, especially poor children. The suggestion is that there is a generational inequity in protecting the elderly poor from the facts and effects of poverty better than we protect poor children. But this suggestion again takes the concern about intergenerational equity out of the broader context of distributive justice. The problem facing poor children has been the vulnerability of programs aimed at poor people: Some administrations have dramatically increased defense budgets at the expense of social welfare spending aimed at meeting minimum requirements of distributive justice. The issue is miscast if it is portrayed as competition between children and the elderly. Rather, programs aimed at distributive justice in general—redistributive transfers from the rich to the poor—have all taken a back seat, except where welfare redistributions to the elderly poor have been protected by the universal or all-class ''savings'' function of Social Security.

Our concern about increased poverty among children and our legitimate worries about the stability of the transfer systems that will soon encounter the baby boom cohort should not tempt us to undermine what is valuable in our collective solution to the age-group problem. At the same time, we must devote adequate resources to meeting the needs of poor children—indeed, of the poor at any age. I am skeptical that dismantling the protection we have afforded the elderly poor will really be followed by more adequate transfers to the remaining poor.

8

Halfway Measures

Ideal Theory and Practical Reforms

The Prudential Lifespan Account of the age-group and birth-cohort problems is part of what is known as *ideal* or general-compliance moral theory. The account is ideal in two ways. First, prudent deliberators are to choose principles for institutions that distribute *fair* shares of basic goods over the lifespan. That is, the principles they choose are based on the assumption that there is already compliance with other, more general principles of distributive justice. Thus we suppose that we are living in an otherwise just society in order to solve the age-group and birth-cohort problems. Second, prudent deliberators assume that there will also be general compliance with the principles they choose as a solution to the age-group problem.

Despite these strong assumptions, ideal moral theory is useful. An ideal theory can be used to criticize existing institutions and to point us in the direction of better alternatives; it offers a vision of a just social arrangement. It does not, however, provide a blueprint for moving from nonideal to ideal moral contexts. And, it does not tell us what compromises of principle are permitted in the effort to make social arrangements more just. Moreover, planners and legislators often operate with fewer degrees of freedom than are allowed philosophers—or our prudent deliberators. Ideal theory supposes that we can reshape basic institutions, but reformers tend to work within much narrower confines. To

139

them, ideal moral theory seems utopian—something unachievable. Even though the ideal theory is intended to describe a feasible social arrangement, it does so only in the special sense that such an arrangement would be stable if we could establish it. Even so, in another, quite practical sense it may not seem feasible. It may prompt us, like the famous Maine farmer, to say: "You can't get there from here."

Though ideal theory generally seems out of reach of planners and public policy makers, we should not allow it to get out of sight. I want to explore this assertion from two angles. First, I want to examine what happens when important reforms of a distributive institution, say Medicare, are undertaken without keeping our eye on justice, at least as an ideal target. Cost-containment measures recently enacted involve rationing some beneficial care, and most of the rationing is aimed at the elderly. What happens when the just distribution of health care is not kept in mind as a goal of reform? Second, I want to look briefly at some measures which might make our health-care system function in a more prudent fashion, at least for some individuals. Broadening the use of advance directives, such as Living Wills, may lead others to treat us in ways we think it prudent to be treated. Making available alternative benefit packages in insurance schemes, including insurance that provides for long-term care, may also make it possible for some individuals to exercise prudence and to shape their local health-care environment in ways they think appropriate. The question is whether such halfway measures can approximate what the appeal to prudence seeks in ideal theory.

Rationing Health Care to the Elderly: Why Saying "No" Is so Hard

In the early 1980s, a number of cost-containment measures were introduced to control health-care costs that were rising at rates far greater than inflation. The main federal hospital cost-containment measure was the introduction of "diagnosis related groups" (DRGs), a flat fee paid for the treatment of Medicare patients falling within certain categories. Many states use these regulations as their sole form of hospital cost-containment. The effect is that the most explicit rationing of health care—though we call it cost-containment—is actually aimed at the elderly. It is a form of rationing by age.

If cost-containment measures, such as Medicare DRGs, involved trimming only unnecessary health-care services out of public budgets, they would pose no moral problems. Instead, such measures lead physicians and hospitals to deny some beneficial care, such as longer hospitalization or more diagnostic tests, to their own patients, that is, at the *micro* level (OTA 1983). Similarly, if the *macro* decision not to disseminate a new medical technology, such as liver transplants, resulted only in the avoidance of waste, then it would pose no moral problem. But when is it morally justifiable to say "no" to beneficial care or to useful technologies? Though this rationing is aimed largely at the elderly, it is not possible age-discrimination that concerns me here (for that issue, recall Chapter 5). I am now asking a more general question.

Justification for saying "no" at both the micro and macro levels is possible. In fact, as I suggested in Chapters 4 and 5, one requirement of justice is that—given resource scarcity—saying "no" to some beneficial services must take place. Here, I want to explain why it is especially difficult to justify saying "no" in the United States, and why, without such justification, cost-containment measures risk luring physicians into violating their moral duties to their patients. Thus, halfway measures or reforms raise special moral problems of their own.

As we have seen, principles of distributive justice are needed to resolve disputes about who should get what when not everyone can get what they want. We would not need principles of justice to resolve disputes about how resources should be distributed if we were in the Garden of Eden. Some suggest that we have operated in our health-care system as if we lived in the Garden of Eden. Retrospectively, fee-for-service reimbursements have made the meeting of patient needs seem cost-free to providers and patients alike. But this suggestion really is an illusion. We have traditionally faced resource scarcity by rationing according to the ability to pay, which I have argued is a morally indefensible way of saying "no." For most of our history, this has meant significant differences in health-care inputs and health-status outputs between socioeconomic groups and between the races. Though Medicare and Medicaid have closed major gaps left by private health-care insurance, as I noted earlier, we still have 25 to 30 million people without health-care insurance.

Some people believe that there is no need to ration beneficial medical services. They claim our problem is imprudent or unjust expenditures elsewhere in the social budget (such as for defense or cosmetics), not real scarcity. They insist sick patients should not bear the brunt of the trade-offs between guns and bandages. This objection has some merit. If scarcity in one area is the result of clear injustice in another, then rationing the scarce resource may only compound the injustice. Still, even if scarcity is the result of our living in a partially unjust world, it is a practical reality we must face, and principles of justice can provide a critical perspective from which to assess our institutions. More generally, health care is only one important social good among others. If it does not compete with defense, then it must do so with education or with other welfare redistributions. Moderate scarcity is real and we should seek appropriate principles of justice.

Because of scarcity and the inevitable limitation of resources even in a wealthy society, justice—however we understand it—will require some "no"-saying at both macro and micro levels of resource allocation. No plausible principles of justice will entitle an individual patient to every potentially beneficial treatment. Providing such treatment might consume resources to which another patient has a greater claim. Similarly, no class of patients is automatically entitled to whatever new technology might offer them some benefit. New technologies have opportunity costs, consuming resources that could be used to produce other benefits, and other classes of patients may have a superior claim that resources be invested in alternative ways.

How rationing works depends on which principles of justice apply to health care. For example, some people believe health care is a commodity or service no more important than any other and that it should be distributed according to ability to pay. For them, saying "no" to patients who cannot afford certain services (quite apart from whether income distribution is itself just or fair) is morally permissible. Indeed, providing those services might seem unfair to patients who are required to pay.

In contrast, other theories of justice—including my own—view health care as a social good of special moral importance. I have argued here that health care derives its moral importance from its effect on opportunity. Specifically, an individual's share of the normal range of opportunities available in his or her society is reduced when disease or disability impairs normal functioning. Since we have social obligations to protect equal opportunity, we also have obligations to provide access, without financial or discriminatory barriers, to services which adequately protect or restore normal functioning. We also must weigh new technologies against alternatives and judge the overall impact of introducing them on equal opportunity. As a result, people are entitled only to those services which are part of a system that on the whole protects equal opportunity. Thus, even an egalitarian theory that holds health care is of special moral importance justifies sometimes saying "no" at both the macro and micro levels.

Saying "No" in the British National Health Service

Aaron and Schwartz (1984) document how beneficial services and technologies have had to be rationed within the British National Health Service, since its austerity budget allows only half the level of expenditures (as a percentage of GNP) made in the United States. The British, for example, use less x-ray film, provide little treatment for metastatic solid tumors, and offer renal dialysis only to the young. Saying "no" takes place at both the macro and micro levels.

Rationing in Great Britain takes place under two constraints which do not operate at all in the United States. First, although the British say "no" to some beneficial care, they nevertheless provide universal access to high quality health care. In contrast, more than 10 percent of the population in the United States lacks insurance. Moreover, race differences in access to health care and health status persist. Second, saying "no" takes place within a regionally centralized budget. Decisions about introducing new technologies involve weighing the net benefits of alternatives within a closed system. When a technology is rationed, it is clear which resources are available for alternative uses. When a technology is widely used, it is clear which resources are unavailable for other uses. No such closed system constrains American decisions about technology dissemination, except—on a small scale and in a derivative way—within some health maintenance organizations.

These two constraints are crucial to justifying British rationing. The British practitioner who follows standard practice within the system does not order the more elaborate x-ray diagnosis that might be typical in the United States, possibly even despite the knowledge that additional information would be useful. Denying care can be justified as follows: "Though my patient might benefit from the extra service, ordering it would be unfair to other patients in the system. The system provides equitable access to a full array of services which are fairly allocated according to professional judgments about which needs are most important to meet. My patient is not entitled to the extra treatment.'' The salve of this rationale may not be what the practioner uses to ease qualms about denying beneficial treatment but it is available.

A similar rationale is available at the macro level. If British planners believe alternative uses of resources will produce a better set of health outcomes than introducing coronary bypass surgery on a large scale, they will say ''no'' to a beneficial technology. But they have available the following rationale: ''Though we would help one group of patients by introducing this technology, its opportunity cost is too high. We would have to deny other patients services which are more important to provide. Saying 'yes' instead of 'no' would be unjust.''

These justifications for saying ''no'' at both levels have a bearing on physician autonomy and on moral obligations to patients. Within the standards of practice determined by budget ceilings in the system, British practitioners remain autonomous in their clinical decision-making. They are obliged to do the best possible for their patients within those limits. Their clinical judgments are not made ''impure'' by institutional incentives, aimed at enhancing hospital profits or ''surplus revenues,'' which pressure physicians to deny beneficial care.

The claim made here is not that the British National Health Service is just, but that considerations of justice are explicit in its design and in decisions about the allocation of resources. Because justice plays this role, British rationing can be defended on grounds of fairness. Of course, some ''no''-saying, such as denying renal dialysis to elderly patients, may raise difficult questions of justice. (I addressed these in Chapter 5.) The issue here, however, is not the merits of each British decision, but the framework within which they are made.

Saying ''No'' in the United States

Cost-containment measures in the United States reward institutions, and in some cases practitioners, for delivering treatment at a lower cost. Hospitals that deliver treatment at less than the DRG rate are allowed to pocket a percentage of the difference. Hospital administrators therefore scrutinize the utilization decisions of physicians, pressuring some to deny beneficial care. Many physicians cannot always act in their patients' best interests, and they fear worse effects if DRGs are extended to physician charges. In some health maintenance organizations (HMOs) and preferred provider organizations (PPOs), there are financial incentives to the group to shave the costs of treatment, if necessary, by denying some beneficial care. In large HMOs, where risks are widely shared, denial of beneficial care may be no greater than under fee-for-service reimbursement (Yelin et al. 1985). But in some capitation schemes, that is, schemes which give a flat fee to physicians for the care of a patient, the individual practitioner is financially penalized for ordering "extra" diagnostic tests, even if he feels his patient needs them. More ominously, some hospital chains are offering physicians a share of the profits made from the early discharge of Medicare patients.

When economic incentives to physicians lead them to say "no" to beneficial care, there is a direct threat to what may be called the Ethic of Agency. In general, granting a physician considerable autonomy in clinical decision-making is necessary if he is to be effective as an agent pursuing his patients' interests. The Ethic of Agency constrains this autonomy in ways that protect the patient, requiring that clinical decisions be (1) competent; (2) respectful of patient autonomy; (3) respectful of other patient rights, for example, that of confidentiality; (4) free from consideration of the physician's interest; and (5) uninfluenced by judgments about the patient's worth. Incentives that reward physicians for denying beneficial care clearly risk violating constraint (4), which, like (5), is intended to keep clinical decisions "pure," or aimed at the patient's best interest.

Rationing need not violate the constraint that decisions must be free from consideration of the physician's interest. British practitioners are not rewarded financially for saying "no" to their patients. Because our cost-containment schemes give incentives to violate this contraint, however, they threaten the Ethic of Agency. A patient would be foolish to think the physician who benefits from saying "no" is any longer his agent. (Patients in the United States have traditionally had to guard against unnecessary treatments, since retrospective, fee-for-service reimbursement schemes have generally provided incentives to overtreat.)

American physicians face a problem even when the only incentive for denying beneficial care is the hospital's, not theirs. For example, how can they justify sending a Medicare patient home earlier than advisable? Can they, like their British peers, claim that justice requires them to say "no" and that therefore they do no wrong to their patients?

American physicians cannot make this appeal to the justice of saying "no." They have no assurance the resources they "save" will be put to better use elsewhere in the health-care system. Reducing a Medicare expenditure may only mean there is less pressure on public budgets in general, and thus more opportunity to invest the savings in weapons. Even if the savings will be freed for use by other Medicare patients, American physicians have no assurance that the resources will be used to meet the greater needs of other patients. The American health-care system, unlike the British system, establishes no explicit priorities for how resources are to be used. In fact, the savings from saying "no" may be used to invest in a technology that may never return care of comparable importance to what the physician is denying his patient. In a for-profit hospital, the profit made by denying beneficial treatment may be returned to investors; in a nonprofit hospital, "surplus revenues" can be used for many purposes that have no bearing on patient welfare. In many cases, then, the physician can be quite sure that saying "no" to beneficial care will lead to greater harm than providing the care would. This makes it morally hard to deny the care.

Saying "no" in the United States at the macro level faces similar difficulties. A hospital deciding whether or not to introduce a transplant program competes with other medical centers. To remain competitive, its directors will want to introduce the new service. Moreover, they can point to the dramatic benefit the service offers. How can opponents of the transplant program respond? They may (correctly) argue that it will divert resources from other projects—projects that are perhaps less glamorous, visible, and profitable but that nevertheless offer comparable medical benefits to an even larger class of patients. They insist that the opportunity costs of the new technology are too great.

This argument about opportunity costs, so powerful in the British National Health Service, loses its force in the United States. The alternatives to the transplant program may not constitute real options, at least in the climate of incentives that exists in the United States. Imagine someone advising the Humana Hospital Corporation, "Do not invest in artificial hearts, for you could do far more good if you established a prenatal maternal care program in the catchment area of your chain." Even if correct, this appeal to opportunity costs is unlikely to persuade Humana, not because Humana is greedy, but because it responds to the incentives society offers. Artificial hearts, not prenatal maternal care programs, will keep its hospitals on the leading technological edge: If they become popular, they will bring far more lucrative reimbursements than preventing low-birthweight morbidity and mortality would. The for-profit Humana, like many nonprofit organizations, merely responded to existing incentives when it introduced transplant programs during the early 1980s, at the same time prenatal care programs lost their federal funding. Similarly, cost-containment measures in some states led to cutting social and psychological services but left high technology services untouched (Cromwell and Kanak 1982). Unlike their British colleagues, American planners cannot say, "Justice requires that we forego this technology because the resources it requires will be better spent elsewhere in the system. It is fair to say 'no' to this technology because we can thereby provide more important treatments to other patients."

The failure of this justification at both the micro and macro levels in the United States has the same root cause. In our system, saying "no" to beneficial treatments or technologies carries no assurance we are saying "yes" to even more beneficial ones. Our system is not closed; the opportunity costs of a treatment or technology are not kept internal to it. Just as important, the system as a whole is not governed by a principle of distributive justice, appeal to which is made in decisions about disseminating technologies. And our system is not closed under constraints of justice.

Some Consequences

Saying "no" to beneficial treatments or technologies in the United States is *morally* hard because providers cannot appeal to the justice of their denial. In ideally just arrangements, and even in the British system, rationing beneficial care is nevertheless fair to all patients. Cost-containment measures in our system carry with them no such justification.

The absence of this rationale has important effects. It supports the feeling of many physicians that current measures interfere with their duty to act in their patients' best interests. Of course, physicians should not think that duty requires them to reject all resource limitations on patient care. But it is legitimate for physicians to hope they may act as their patients' advocate within the limits allowed by the just distribution of resources. Our cost-containment measures thus frustrate a legitimate expectation about what duty requires. Eroding this sense of duty will have a long-term destabilizing effect.

The absence of a rationale based on justice also affects patients. Resource constraints mean that patients can legitimately expect only the treatments due them under a just or fair distribution of health-care services. But if beneficial treatment is denied even when justice does not require or condone it, then patients have reason to feel aggrieved. Patients will not trust providers who put their own economic gain above meeting patients' needs. They will especially distrust physicians in schemes which allow doctors to profit by denying care. Conflicts between the interests of patients and physicians or hospitals are not a necessary feature of a just system of rationing care. The fact that such conflicts are central in our system will make patients suspect there is no one to be trusted as their agent. In the absence of a concern for just distribution, our cost-containment measures may make patients seek the quite different justice of tort litigation, further destabilizing the system.

Finally, these effects point to a deeper issue. Economic incentives such as those embedded in current cost-containment measures are not a substitute for *social* decisions about health-care priorities and the just design of health-care institutions. These incentives to providers, even if they do eliminate some unnecessary medical services, will not ensure that we meet the needs of our aging population over the next several decades in a morally acceptable fashion or that we will make effective—and just—use of new technologies. These hard choices must be faced publicly and explicitly.

Ideal theories of justice, then, do not merely provide a basis for critiques of the existing system. If the pursuit of justice is not our goal when we undertake halfway measures, even those intended to address such serious problems as cost-containment, then our reforms will create serious moral problems of their own. In this case, we risk undermining the Ethic of Agency that governs physicians because we have not paid attention to matters of justice. A similar point was noted in Chapters 2 and 6: In the absence of just social arrangements, other moral obligations, such as filial obligations, are strained and their limits are made unclear. Ideal justice may be out of reach but it must never be out of sight.

Individual Prudence and Self-binding

Some people may insist that we live in a cruel and unfair world. Specifically, they may believe that the prudential planning of social institutions, of the sort prescribed in the Prudential Lifespan Account, is not politically feasible—at least not for us now. Can we undertake halfway measures that at least mimic and perhaps approximate what ideal theory recommends? If we encourage *individuals* to plan prudentially for their whole lives, for example, can we approximate ideal *social* outcomes?

In Chapter 3, I argued against using the choices made by fully informed, young individuals as a basis for the Prudential Lifespan Account. I was concerned that giving too much weight to prudent choices by the young may leave the old at risk. But now we are not seeking to characterize the ideal. We ask instead: If ideal prudential planning at a social level is not politically feasible now, then is individual prudential planning an acceptable halfway measure? Notice that this form of "privatizing" is not under discussion as the result of moral objections to the model of social prudence I have developed. The "privatizing" strategy does not constitute an alternative form of social justice; it only approximates it.

A variety of devices and schemes fall under the heading of individual lifespan planning. One strategy involves *advance directives* through which we instruct others how we want to be treated in certain contingencies. These are especially important if what we view as prudent is unlikely to match what existing institutions might do to or for us. Living Wills are a well-known example of such prudential planning. Such wills instruct third-parties, such as hospitals and physicians, that we do not want certain kinds of medical interventions, such as resuscitation, when we fall under a specified set of medical conditions. In general, we aim to protect ourselves against others deciding what is in our interest when we are in no condition—because of loss of consciousness or competency—to instruct them how to take care of our interests.

Usually, such measures are thought of as promoting individual autonomy: Through them we extend our ability to consent to or refuse actions by others, even though we are in a state in which actual consent or refusal is impossible to obtain. But if these measures are thought out carefully and prospectively, in protecting autonomy they also create a local environment around us which is prudentially planned. Moreover, the attention required to construct such a Living Will, or other advance directives, itself stimulates prudential planning. We will pay attention to what it is prudent for us to want over the lifespan and instruct others that our prudent choices are to be respected.

A Living Will requires social cooperation, however. As an individual declaration it is pointless; others must recognize its force for it to have prudential effects. At the extreme, if we want to promote this form of individual prudence socially, we might *require* individuals to construct such wills, or at least give them strong incentives to do so. Other forms of prudential planning over the lifespan involve even more elaborate forms of social cooperation. If insurance schemes were available that provided protection against long-term-care needs, and not just against the need for acute-medical care, then individuals would have an instrument that enhanced their ability to make prudential choices over the lifespan. Such insurance schemes, for example, might involve trading some acute-care benefits for long-term-care benefits, to keep the price of the premium within range. To be feasible, such plans might have to be purchased at a relatively early point in life. Insurance schemes such as these could mimic, for participating individuals, the effects of a universal, social scheme of the sort recommended by the Prudential Lifespan Account.

Individual prudential planning for income support over the lifespan could also be enhanced for many individuals now excluded from such planning. Many workers, for example, acquire no vested assets in pension plans unless they have worked for some relatively long period for the same company. Many contributions to pension plans are nontransferable to other plans. Consequently, individuals whose work is unstable, because of periods of layoffs or because many lateral moves among employers are a feature of the industry, cannot engage in adequate prudential planning of this sort. Proposals to reform pension plans through Federal regulation, so that a greater number of workers can carry their pensions around with them, would promote more widespread prudential planning by individuals.

No doubt there are many other structures we could devise which would encourage individuals to plan prudently over their lifespans. Individuals who did so would produce mini-environments surrounding themselves in which distributions over their lifespan were prudently allocated—at least from the perspective of their initial plan. Surely these individuals will be better off if distributions to them are prudently allocated. Better some than none; better more than fewer. Reforms that result in such improvements, even for some people, would thus seem genuine steps in the direction of making distributions overall more just.

Although some reform is surely better than none, I shall argue for some caveats. Halfway measures bring with them moral issues that must be addressed. Some of these issues will affect choices among possible reforms of this sort.

The first caveat is that many people will be left out of reform measures if they are of the halfway variety. I have argued that there are social obligations to distribute goods justly—and prudently—between age groups. But individual measures of the sort we are considering will not be universal. What is worse, the groups that tend to be excluded from these reforms will be the economically worst-off groups in society. Lower levels of disposable income and of education would mean that few below the upper-middle class would undertake these steps. Given the distribution of poverty and education in our society, minority groups will also be disproportionately left out. The effect may be to increase inequalities in ways that violate more general requirements of justice.

A second caveat is that the people who are left out of these reforms may be groups who are politically least able to pursue their inclusion in them. If there is a chance to reform a system to make it more just overall, it may require the united efforts of various social groups, all of whom may benefit. If we pursue schemes that promote prudence for the best-off segments of society, then we may undermine the ability of the worst-off to enjoy similar benefits in the long run. Together, the first and second caveats force us to think carefully about whether some reform is always better than none and about the strategies which in the long run will most promote justice for all. But these caveats are quite general, raising moral and political issues that arise regarding any reforms. More specific and more interesting issues are raised by some additional remaining caveats.

The third caveat is that reforms promoting individual prudential planning over the lifespan may lead to consequences that many elderly think are far from ideal. The point here is simply the practical implication of the theoretical problem raised in Chapter 3: Prudential planning by young individuals may be biased against what the elderly take to be prudent. My model for *socially* prudent choices abstracted from the biases of the young or the old (by means of the "veil of ignorance" and the index of primary social goods). In these reforms we must live with these biases. The general point is that the young may misestimate what their preferences and values late in life will be, and their choices will not be ideally prudent as a result. Binding ourselves to such imprudent choices—in the name of prudence—would no doubt produce regrets and a desire to alter prior commitments. Of course, what can be said in favor of this procedure is that we are stuck with the results of *our own* choices and *our own* mistakes. Having to live with the results of our errors is part of the human condition, and it is especially part of that condition if we promote autonomy. Nevertheless, we should understand that we are only approximating what we would achieve if we could bind ourselves to ideally prudent allocations: Our regrets about this approximation may fall at a point in life when we can least revise our plans.

A fourth caveat concerns the problems produced by changing preferences. There are two distinct cases to be considered here. We are not perfectly rational creatures. We can anticipate situations in which our will is certain to be weak and where we will want to do what we know is imprudent (cf. Elster 1984). When Ulysses stuffed wax in his sailors' ears, had them bind him to the mast, and instructed them to ignore his pleas to be released, he was protecting himself against the irrationality he knew the sirens could induce. The first kind of case, for which self-binding prudential schemes seem morally unproblematic, are those in which we are guarding against our own future irrationality.[1] Third parties in these cases find their task less problematic if they too can be sure that the preferences they are to ignore are really irrational or the results of an "inauthentic" part of ourselves (see Schelling 1986).

Not all self-binding schemes are so unproblematic, which brings us to the second kind of case. Sometimes there are endogenous changes in our preferences which by no means seem irrational or unreasonable, at least considered from the point in our lives at which they arise. If Ulysses had allowed himself to be lured by the sirens, he would have destroyed the project of greatest importance to him. Consider a person who issued advance directives that he not be given heroic medical measures if he found himself in certain conditions, and suppose he used the money that would have been spent on such measures to insure himself against other kinds of risks or to obtain a good education. Later in life, facing these conditions, he thinks a few weeks more of life is

---

1. See Elster (1984) for a definition of "self-binding."

worth everything to him. It is hard to insist that this revision in his views is now irrational. Even if it would have been irrational for him to forego that early education in order to save money for the heroic measures, living his whole life according to that contingency, it is not irrational for him to now want the measures. It is not "locally" irrational, at this point in his life, even if it would have been prospectively, when he thought globally about his lifetime well-being.[2]

Third parties called upon to enforce self-binding plans, in these cases, may find themselves facing hard moral choices. In the case of the revised preference for heroic measures, third parties may still be able to appeal to a more global notion of rationality for the person in question. They may hold him to his earlier choice. But not all cases of endogenous change of preferences (or values) can be dismissed as cases where an appeal to global rationality will override local rationality. Sometimes, a change in preference represents what would have been part of a coherent, rational plan of life: It is just not the kind of change anticipated in the earlier, self-binding plan. In such a case, third parties may feel that an earlier self has imposed a conception of what is good on a later stage or self. Such an imposition may itself be imprudent. If, as third-parties, they enforce the earlier choice, they will be acting inappropriately. What would have been justifiable paternalism—an intervention authorized by a self-binding plan that appropriately anticipated irrationality—now becomes morally problematic interference with the autonomy of the later self. Cases like this, as Schelling (1986) notes, raise very difficult moral issues.

Some of the complexity of these cases is brought out by the warning we sometimes hear when we have not seen someone for a long time— "Don't be surprised, he's a different person now!" Over long periods of time, a person may have changed many dispositions, values, character traits, and tastes—he may not even "identify" with his former self. Self-binding arrangements made by an earlier self thus may well appear as unjustifiable intrusions into the life of the later self. Parfit (1984) has argued that the expression, "He's a different person now!" may be literally and not just metaphorically true. Personal identity, he argues, depends on facts about physical and psychological continuity and connectedness, not on some further fact, like a Cartesian ego, which remains the same through all qualitative changes. If he is right,

---

2. I am indebted to Dan Wikler for the example and the term.

then a systematic problem of paternalism faces any attempt at self-binding. (I discuss Parfit's views in the Appendix.) Regardless of how defensible Parfit's metaphysical claim is, third parties will face an issue that has most of the features of problems about paternalism.

The fourth caveat comes to this: What kinds of self-binding devices are morally permissible is an interesting and complex moral question. We must consider what kinds of revocability (under what sorts of conditions) are needed in order to prevent self-binding plans from binding others (including our later selves) to do things that violate autonomy. Notice that in ''privatizing'' prudence, in order to approximate ideally just arrangements, we have raised further problems of *justice*. Are the devices we need to protect ourselves against imperfect rationality compatible with our beliefs in autonomy? Despite the great intrinsic interest of these problems raised by advance directives, I must here stop short of discussing them. They are the topic for another book.

A Final Word

The patient reader, no doubt with some alarm, has watched a simple intuition about the distinctive nature of the age-group problem grow into a baroque construction. The Prudential Lifespan Account is replete with frames, information restraints, and assumptions about general compliance. Applications of the account throughout the book have been highly qualified. And in this chapter the reader has been reminded how difficult it is to use theory to guide practice—but also how difficult it is not to do so. An exasperated physician at a conference once rebuked me, ''Philosophers are complexifiers—you can't get a straight answer out of them. It's lucky they are not doctors!'' Perhaps it is lucky they are not planners or legislators either.

Despite this mea culpa, I do not want the reader to lose sight of my central message. To simplify it, think of it as a heuristic approach, not a theory. As a society, we are harrangued by Cassandras who see the aging of society as a divisive problem. They see competition, with the old pitted against the young, grandparent against grandchild, in a life and death scramble for scarce resources. By focusing on what is distinctive about age—that is, we all age—I have found a way to undercut this grim view.

I offer a unifying vision. We all pass through institutions that distribute goods over our lifespan. If these institutions are prudently designed, we *each* benefit throughout our lives. It is only prudent to treat ourselves differently at different stages of life, as our needs change. What is prudent with respect to different stages of a life determines what is fair between age groups. Prudence here guides justice. If as policy makers, planners, and the general public we can all keep our eyes on this unifying vision, and if we can ignore the divisive talk about competition, then our target will be policies that benefit us all over our whole lives. Establishing such policies would mean doing justice to the old and the young.

Appendix

Problems with Prudence

Challenges to the Classical Model

The Prudential Lifespan Account rests on two assumptions which I have treated so far as uncontroversial. The first assumption is that there is an important difference between distributive problems that cross the boundaries between persons, on the one hand, and problems that involve allocation within a life (between the stages of a life), on the other. The elaborate "frame" I constructed to isolate the age-group problem from interpersonal transfers of goods is motivated by the view that persons are basic entities for purposes of the theory of justice. For example, taking goods from one person to benefit another requires justification. More generally, deciding which inequalities between persons are justifiable is the central issue in distributive justice. By showing that the problem of distribution between the young and the old, which appears to be interpersonal, can be reduced to a problem of intrapersonal, prudential allocation, I try to replace a more complex and problematic issue with a less problematic one. If the boundaries between persons are not so important, or if similar problems arise for distributions within a life as between them, then the motivation for my strategy is undermined.

The second assumption that I have treated as uncontroversial is this:

Rationality requires that we show equal concern for all parts of our (future) lives. Parfit (1984:313) refers to this assumption as The Requirement of Equal Concern. I used this assumption in Chapter 3, to justify the restrictions placed on prudent deliberators, and in later chapters it affected the reasoning of prudent deliberators choosing how to distribute health care and income support over the lifespan. If rationality does not require that we be equally concerned about all parts of our future, then my arguments will have to be qualified and my account will turn out to be based on stronger presuppositions than it now appears to have.

These two assumptions are quite standard in moral philosophy. The assumption about the importance of the boundaries between persons appears in diverse normative theories, ranging from Nozick's (1974) libertarianism to Rawls's (1971) justice as fairness. It also motivates, at least in part, the appeal of contractarian approaches to the foundations of ethics, for contracts respect the distinctness of persons. Assuming The Requirement of Equal Concern is no less standard, both in philosophy and in economic theory. Nevertheless, both of these assumptions have come under powerful criticism in Parfit's (1984) brilliant and provocative work, *Reasons and Persons*. My task in this Appendix, which will be of interest primarily to those with philosophical training or interests, is to provide some defense of these assumptions and to offer qualifications, where necessary, in the face of Parfit's arguments.

Rationality and the Requirement of Equal Concern

We might begin by noting some features of the classical theory of individual rationality, which Parfit calls the Self-interest Theory, or SI. The central claim of this theory is

(S1) For each person, there is one supremely rational ultimate aim: that his life go, for him, as well as possible (Parfit 1984:4).

Other claims of the theory are

(S2) What each of us has most reason to do is whatever would be best for himself, and

(S3) It is irrational for anyone to do what he believes will be worse for himself, and

(S4) What it would be rational for anyone to do is what will bring him the greatest expected benefit (Parfit 1984:8).

SI involves other claims, but they are not relevant to the argument here. In his subsequent discussion of SI (in Part Three of his book), Parfit says it is also committed to

The Requirement of Equal Concern: A rational person should be *equally* concerned about *all* the parts of his future.

The Requirement of Equal Concern, we may take him to mean, is implicit in the other claims of SI. For example, the greatest expected benefit (in S4) is aggregated over the whole life. In some classical formulations of the theory, such as Sidgwick's (1907) and Rawls's (1971), The Requirement of Equal Concern is explicit.

Parfit considers a variety of arguments against SI. In Part One of his book, he considers arguments that SI is self-defeating. In Part Two, he argues that SI occupies an untenable and arbitrary position in between morality and an alternative account of individual rationality, namely, the (Critical) Present Aim Theory. Morality is neutral about time and persons. It takes interests and desires into account regardless of whose life they involve and regardless of when they occur within a life. In contrast, the Present Aim Theory is biased in favor of one's own present aims, and so is nonneutral with regard to both time and persons. The Self-interest Theory is neutral about time but biased about persons. Parfit argues that this in-between position is untenable because there is no consistent basis for being neutral with regard to time but not persons. These are important arguments, which have been discussed elsewhere (see Kagan 1986). If we think that the way in which experience involves persons through time gives some reason to be neutral about one but not the other, then these arguments will not be persuasive. I want to concentrate in any case on Parfit's ultimate challenge to the importance of persons.

In what follows, shall consider only those of Parfit's arguments against SI that rest on claims about personal identity. Parfit suggests that it is uncritical, common sense acceptance of a Non-Reductionist View, for example, the belief in a Cartesian ego or mental substance,

which leads us to think that rationality implies The Requirement of
Equal Concern. On such a view, since the ego or mental substance is
what makes each of us the person we are over time, then we think we
must be equally concerned about what happens to "it" at all points.
There is what Parfit calls a "further fact" over and above the psychological continuities. In contrast, on Parfit's Reductionist View, the facts
relevant to deciding whether we are the same person over time are facts
about the connectedness and continuity of mental events (and perhaps
some facts about their causes). If these are the underlying facts, however, then the metaphysical basis for The Requirement of Equal Concern may be removed. Since psychological connectedness and continuity can vary in degree, the door may be open for us to care more
about some parts of our lives than others. To understand Parfit's claims,
we must look more closely at his terminology and arguments.

Parfit (1984:211) says that on the Reductionist View, "each person's
existence just consists in the existence of a brain and body, and the
occurrence of a series of interrelated physical and mental events." The
central features of the position are

1. that the fact of a person's identity over time just consists in the
   holding of certain more particular facts,
2. that these facts can be described without either presupposing the
   identity of this person, or explicity claiming that the experience in this
   person's life are had by this person, or even explicitly claiming that
   this person exists. These facts can be described in an *impersonal* way.

If we reject either or both of (1) or (2), we get a Non-Reductionist
view. A Non-Reductionist holds that "personal identity over time does
not just consist in physical and/or psychological continuity. It is a
separate, further fact," for example, a fact about the persistence of a
purely mental entity such as the Cartesian ego (Parfit 1984:210).

Through a series of thought experiments, involving teletransporting
replicas of persons to distant places (in the way Mr. Spock may be
"beamed" down to a planet in *Star Trek*), or replacing parts of one
person's brain with parts of another, Parfit argues that only two kinds of
facts can and should underlie our judgments about personal identity.
Suppose we want to know whether two "person stages," A and B, are
stages of the same person. If A and B share desires, beliefs, and intentions, or if B remembers having an experience A had, or if B carries out

A's intentions, then A and B have *direct psychological connections*. If enough direct connections hold between B and A they are *strongly connected*. When overlapping chains of strong connectedness hold between person stages, then they have *psychological continuity*. Parfit's main arguments in Part Three are intended to show that only facts about psychological connectedness and continuity are relevant to answering questions about personal identity. I shall return later to consider some further points about the concept of personal identity and Parfit's metaphysics, but for now let us consider how Parfit uses these metaphysical conclusions.

Parfit first argues that, if one is a Reductionist, an Extreme View is not irrational. On this Extreme View, first suggested in Butler's criticism of Locke and echoed in Sidgwick (cf. Parfit 1984:307), we seem to have no reason to be concerned about later stages of ourselves if no further fact, beyond connectedness and continuity, links us to them. If personal identity does not involve a deep further fact, namely, about a persistent Cartesian ego or mental substance, it is a less deep fact or involves less (Parfit 1984:312). We may then have no more reason to care about later stages of ourselves than we do about other persons.

But a Moderate View is also not irrational, and it is this view which most interests us. Parfit argues that we should each then reason as follows: My reasons for caring about my future, if I care at all, must depend on my concerns about my connectedness and continuity (with or without normal causes) with my later stages. But then, it may be rational to care less about parts of my future with which I am less strongly connected. That is, I can "discount" or care less about the well-being of person stages less strongly connected to me than I do about those with which I am more strongly connected. On this view, it is not the mere *temporal* remoteness of future stages that leads to my caring less. Here there is no disagreement with the classical Self-interest theorist who condemns what economists call "pure time preferences." Parfit argues that since connectedness is a relation that holds to a lesser degree with future stages, then it is defensible to believe it has a different degree of importance. Therefore it is not irrational for me to discount the well-being of my future person stages according to their degree of connectedness with me now.

This conclusion, Parfit argues, is incompatible with the Self-interest Theory, since it seems to involve a rejection of The Requirement of Equal Concern. Instead of being required to care equally about all parts of my future, I may care less about parts of it to which I am less connected. Parfit says we are not rationally *required* to have this discount rate, only that it is not irrational to have it.[1] Revising the Self-interest Theory to include a discount based on degree of connectedness would destroy the central claims of the theory, however. Parfit (1984:317) says "it would break the link between the Self-interest Theory and what is in one's own best interests." That is, it contradicts S3, the claim that it is irrational for anyone to do what he believes will be worse for him. "If it is not irrational to care less about some parts of one's future," Parfit (1984:317) concludes, "it may not be irrational to do what one believes will be worse for oneself. It may not be irrational to act, knowingly, against one's own self-interest."

Parfit's argument is more puzzling on close examination than it first appears to be. There may be something close to an equivocation involved in it. What are one's "own best interests" or "one's own self-interest"?[2] Are the future interests I am discounting really my own interests? To make his point, Parfit requires that these interests be one's own. These discounted interests of future person stages must count as interests of the person doing the discounting. If I discount them, they must still be *mine* to discount. Though they are really mine, I care about them less.

If, however, personal identity is determined by the fact that there exists between two person stages an appropriate degree of connectedness and/or continuity (with the right kind of cause), that is "Relation R" holds (as Parfit 1984:215 refers to it), then perhaps this later stage, whose interest I am discounting, is not fully or completely part of me. Its interests are not really or fully (or to an adequate degree) *mine*. But then, if I care less about *those* interests, I am not caring less about *my* interests. Consequently, I am not necessarily doing what is worse for myself if I ignore what are not completely or really my interests in any case. That is, if personal identity is a matter of degree, resting solely on

---

1. I am not sure why we are not required to have the rate on the view that the sole grounds for caring about our future is our concern about connectedness and continuity.

2. Parfit's argument does not turn on a particular equivocation about "interests." Suppose interests were defined independently of a person's preferences or what he cares about, and that we took prudence to involve the maximal satisfaction of interests over a lifespan. Then if rational action involved the attempt to satisfy ones preferences or aims, it would indeed be problematic whether prudence was required by rationality. This argument does not turn on any points about personal identity and is not Parfit's.

facts about psychological connectedness and continuity, then the following is true: If I discount the interests of relatively unconnected stages of myself, I am in the relevant sense still attending only to those interests which are really mine.[3]

On this reading of Parfit's claims, I make personal identity depend on degree of connectedness and continuity, in the same way I make my concern so dependent. But this means I have not really rejected The Requirement of Equal Concern. I still show my dominant concern for all and only what is me—that is all I care about. The Revised Self-interest Theory, which includes the discount rate for connectedness, only appears to be discounting *my* future well-being. Actually, it is just identifying (*isolating,* more precisely) what counts as *my* future well-being. The dispute between the Classical and the Revised Theories, it turns out, is only a dispute about what counts as my future, not a dispute about whether it is *rational* to be concerned about all parts of my future. (Both formulations of the dispute raise problems for my account of how to solve the age-group problem, for both make it harder to justify certain transfers from the young; that is why I must return to the claims about the importance of personal identity.)

In suggesting that Parfit may be equivocating about personal identity here, I may be too contentious (or even mistaken). In any case, I shall return later in the Appendix to discuss further Parfit's views about personal identity. In the remainder of this section I want to consider a less contentious way to express my concerns about Parfit's argument for a discount rate based on the degree of connectedness. I shall be concerned more with what counts as a person's *interests* and less with what counts as the *same person's* interests.

Suppose we formulate corresponding principles of individual rational action for the Classical and the Revised Theories. For the Classical Theory the principle would presumably be: Act so as to maximize personal expected utility over your lifetime. We can think of an individual as having a utility function. Such a function has as its inputs or "arguments" the individual's resources, talents, skills, preferences, and values, and it has as its output particular values of the quantity utility. The Classical Action Principle implies The Requirement of Equal Concern for the following reason: We do not discount *utilities* or

---

3. Of course, as Parfit (1984:206) notes, strong connectedness cannot be a criterion of personal identity because it is not transitive.

well-being merely because of the time at which the utility occurs or because of the degree of psychological connectedness of the person having the utility with other stages of himself. We discount only for the reduced *probability* of future utilities (this is what "expected utility" means).

What does the corresponding action principle for the Revised Self-interest Theory look like? To answer this question, we must first understand how to discount *classical utilities*. A discount rate based on psychological connectedness may be thought of as a more complex utility function than the one used in the Classical Theory. On this view, we first calculate utility as in the Classical Theory: We then take that result as an argument in a new utility function which includes the discount rate. (Let us ignore the fact that psychological connectedness is not a uniform function over time.) The values (or outputs) of this function will be discounted utility. This discounted utility represents well-being from the perspective of an individual Discounter prospectively assessing the outcomes *he cares about*. The discount rate thus works the way an additional set of preferences would, were they added into the original (Classical) utility function. We can even think of discounted utility as plain utility or well-being *from the point of view of a Discounter*. It is what that Discounter takes to be relevant to his well-being.

Having done the necessary discounting, we are ready to state a principle of rational action for the Revised Theory. A likely candidate is this: Act so as to maximize personal expected discounted utility over your lifetime. Suppose a Discounter adopts this principle. He now can comply with the basic insight of the Classical Theory, that it is irrational for anyone to do what he believes will be worse for him. It would certainly be worse for a Discounter to maximize utility over his lifespan rather than discounted utility, since what he prospectively cares about is discounted utility. Indeed, if his utility function includes a discount rate for psychological connectedness—this is what he *cares* about—it would be imprudent for him to maximize utility rather than discounted utility. The Discounter who prospectively acts rationally will make things *worse* for himself only if we take his *interests* or well-being to be captured by utility and not discounted utility. (Note the contrast with my more contentious point. Earlier I complained that net personal expected utility does not capture the *Discounter's* interests because some of that utility does not belong to person stages which are (fully? actually?

completely?) part of *him*. Here the point is that utility does not measure *interests* for the Discounter: only *discounted utility* does.)

The Discounter may seem to be acting in accord with his own best interests only because we have interpreted discounted utility in a particular way. We have treated discounted utility as if *it,* rather than utility, were actually a measure of well-being from the perspective of the Discounter. Perhaps the Discounter should think of discounted utility not as the relevant measure of utility, but as a *distortion* of utility, a distortion induced by the peculiarly curved lens of the discount rate. For him discounted utility is "utility as it matters to me now." He knows (or should realize) that well-being for less-connected stages of himself will really be less if he maximizes his discounted utility. It is just that he cares less about the fact that less-connected stages of himself will be worse-off than they would be if he cared about utility and not just discounted utility.

But this attitude is not to be confused with short-sightedness, which it would be if the Classical Theory were true. If he really believes the discount rate is reasonable, the Discounter will think of the distortion as a *correction*. What is valuable to him (prospectively) is not his utility but his discounted utility. His principle of action is: Act so as to maximize expected discounted utility over your lifespan. In acting on this principle, he will knowingly be acting against his self-interest as measured by expected utility over his lifespan. But since he (prospectively) cares only about discounted utility, he will not be knowingly acting against what in fact he takes to be his own best interests. He will still not be knowingly doing what he thinks is worse for himself.

Parfit's (1984:317) claim is that the Revised Theory, which includes a discount rate for connectedness, "breaks the link between the Self-interest Theory and what is in one's own best interests." My earlier complaint was that this claim is false if we redefine personal identity, and thus what counts as *"one's own* best interests." My second complaint is that if we construe "interests" as a Discounter must, then he will *not* be doing what he believes is worse for himself when he discounts (later) utility. From his perspective, it is not utility, but discounted utility, that defines his lifetime *interests*. (Remember that interests cannot be defined entirely independently of the person's aims and preferences; cf. note 2.)

This interpretation of what is involved in the Revised Self-interest

Theory also involves accusing Parfit of an equivocation. This time the equivocation is on the notion of the same person's *interests* rather than on the notion of the *same person's* interests. Though I believe this equivocation is important, in what follows I shall drop, for the sake of argument, these complaints about equivocation. Instead, I want to point out some problems that arise for Discounters on the assumption that they really are ignoring some of their own interests.

Suppose our Discounter plans his life at an early time E and then ages in the normal way. He now finds himself living at a later stage L of his life which is less connected with E, and his utility level is lower than it would have been had his plan not prospectively discounted utility. From the standpoint of L, E now seems to be the stage of life whose utility the Discounter would prefer to discount. Could he now (at L) retrospectively plan his life, he would discount utility at E, saving more resources for use at L. At L, the Discounter now cares about his utility at L. Looking through the lens of his discount rate, he is not discounting his utility at L, but only his utility at other, less connected points in his life. How he used to care about his utility at L, when he discounted it at E, is not how he cares about it now. Similarly, how he now cares about the utility he had at E is not how he cared about it at E. But the Discounter cannot remake history.

Nor, does it seem, can he learn much from it. At L, he might bemoan his low utility at L, wishing perhaps that his earlier self had not been a Discounter, though he cannot criticize himself for any irrationality unless he is ready to conclude that discounting is irrational. As an inveterate Discounter, however, he has to plan the rest of his life in a way that virtually guarantees that later in life he will again regret his current plans just as he now regrets his earlier ones. Here we have broken the link between acting rationally and improving one's strategies for planing by learning from cases in which we do not do well.

A small point about regret. Parfit (1984:187) remarks that someone who has a bias toward the near future may regret having had the bias in the past, but that gives him no reason to regret still having it. He still has no reason to care that a future, less-connected stage of himself will suffer in the future. We might note, however, that the Discounter may regret his imprudence more than the Classical Non-Discounter. The imprudent Non-Discounter can at least look back at all the utility enjoyed early in life through a lens that does not discount the earlier utility. He can say, "At least I had it good then!" But the Discounter, looking back at an earlier time, when he was not well-connected to his current self, will not care that he enjoyed himself so much then. He cannot console himself by recalling or imagining the full value of his earlier personal utility.

I do not want to beg any questions. The Discounter does well if, at E, he acts successfully to maximize personal expected discounted utility, even if this means that at L he is not as well-off as he would be had he earlier maximized personal utility. At L, however, the Discounter is no longer discounting his utility at L. Judged at L, the results of his "successful" plan at E leave him not only with lower utility but with lower discounted utility. The function that takes utility and the discount rate as its arguments and has discounted utility as its values is one the Discounter uses to assess his well-being *as he ages*. His former successes are responsible for his current dissatisfactions. But if he adheres to the distorting lens of his discount rate, he cannot improve on his dissatisfying outcomes—as judged by himself when those outcomes are realized. Since he prospectively cares less about those later outcomes, which affect less-connected stages of himself, he adheres to his discount rate. But when he eventually becomes that stage of himself, he will be dissatisfied with the results.

When we look at the graph of a continuous function through a magnifying glass, the segment in focus seems discontinuous with the rest of the line. It is on a new scale. We can move the glass along the line, and the effect is not a new, continuous function, but relocation of the discontinuity. A Discounter looks at his well-being through such a magnifying glass. As he ages, the utility he cares about, discounted utility, will lead to different assessments of his well-being for *the same point* in his life. At E, his discounted utility for point L will be a different value from what his discounted utility for L will be at L.[4]

This analogy suggests that adopting a discount rate based on connectedness introduces problems similar to the problems that arise from time-indexed desires. When Brandt (1979) discusses this issue, he concludes that desire-satisfaction accounts of well-being are unacceptable because there is no whole-life perspective from which one can con-

---

4. Whether the Discounter is committed to what have been called inconsistent time preferences (see Elster 1984) is an interesting question. Discount rates that give absolute priority to stages very strongly connected with the current stage, but that give equal and smaller rates to less connected stages, will generate inconsistent time preferences. I ignore this complication for Parfit here.

sistently construct a plan of life that maximizes desire-satisfaction. Instead, he suggests that an "enjoyment" account of happiness can avoid this sort of incoherence. The Discounter avows, as a matter of principle, as it were, that the perspective from which to plan a life is one's present interests, including one's preferred discount rate. But such a person must make assessments of his well-being at many points in his life and he must make new plans as well. He gives himself no concept—such as utility—on the basis of which he can arrive at uniform evaluations of his well-being for the same point in his life. From different points in his life, the Discounter will assign the same point in his life different values of discounted utility. In this sense, there is no way for evaluations and plans to cohere over a lifetime.

When we give up The Requirement of Equal Concern and adopt, instead, a discount rate for utilities based on psychological connectedness, we are abandoning the requirement that there be a uniform way of assessing the well-being of a given point or stage in a life. By making that assessment relative to the point in life at which the assessment is made, we lose the possibility that evaluations and plans will cohere over a lifetime. (It converts the problem of interpersonal comparisons of utility into an intrapersonal problem.) This may seem only a way of spelling out the implications of Reductionism. It may seem only a redescription of what happens when we decide that Relation R (including connectedness) is what matters, not personal identity. But I am inclined to think such *relativity* in assessment, which leaves us without a kind of coherence, may count as an argument against a discount rate of this type. Perhaps we have found a consideration which shows that the fact of reduced connectedness is not enough reason for concluding that connectedness is less important (cf. Parfit 1984:314).

One last point concerns where Parfit's criticisms of the Classical Theory leaves him strategically, given his adherence to some form of Consequentialism (perhaps Utilitarianism). When the Discounter assesses his well-being for purposes of action, either prospectively or retrospectively, he uses discounted utility, not utility. He is thus committed to the following kind of relativism: He will assign different values to his well-being at a given stage depending on how connected he is to that stage at the time he makes the assessment. The measure of well-being for him is thus *local*. The utilitarian, in contrast, places great importance on the fact that his principle of action (Act so as to maximize total expected utility) uses a *universal* measure of well-being for purposes of guiding action. Parfit's strategy for attacking the Self-interest theorist was to show he was arbitrary in splitting his own utility off from everyone else's, caring only about personal utility and discounting the utility of others. The Self-interest theorist tries to adhere to an untenable middle ground that partially, but not completely, localizes utility assessment. The Discounter, in contrast, fully localizes it (at least with respect to connectedness, if not with respect to time). In this sense he is less arbitrary.

But now it seems that an even greater *gulf* has emerged between the concerns of rationality (to do what is best for oneself, using a local measure of well-being), and the concerns of Consequentialism or Utilitarianism, which are committed to universal measures of well-being. What seemed initially plausible (to me, at least) about the Classical Theory was that it appeared to be committed to a nonlocal measure of well-being (no discounting of utilities by time or connectedness). In this regard it resembled Utilitarianism. Both theories could accept that the principle of individual rational action was one requiring the maximization of personal expected utility, a temporally nonlocal measure of well-being. To get from individual rationality to morality it was not necessary to introduce or justify the idea of nonlocality for measures of well-being. It was necessary only to justify the move from one nonlocal measure to another, and this move could be understood as the move from rationality to morality (no wonder it was hard!).

But now the move from rationality to morality is also the move from locality to universality of measures of well-being, which is why the gulf between the two seems so much larger. Parfit pushes individual rationality much further away from morality than we though it was. We might have thought that one attractive feature of utilitarianism was the nonlocality of its measure of well-being, and that this was attractive because it had some connection to how it was rational to carry out assessments of personal well-being. But now all (!) that can be said for utility is that such universality is appropriate from a moral point of view.

Personal Identity and the Boundary Between Persons

In Part Three of his book, Parfit argues that personal identity is a less deep metaphysical fact than our common sense beliefs presuppose it is. Because it is a less deep fact, it may be less important, for example, to moral theory or to the theory of individual rationality. Of greatest concern to us, the boundaries between persons may not be as morally important as we make them out to be in the theory of distributive justice. As Parfit (1984:339) puts it,

> If some unity is less deep, so is the corresponding disunity. The fact that we live different lives is the fact that we are not the same person. If the fact of personal identity is less deep, so is the fact of non-identity. There are not two different facts here, one of which is less deep on the Reductionist View, while the other remains as deep. There is merely one fact, and this fact's denial. The separateness of persons is the denial that we are all the same person. If the fact of personal identity is less deep, so is this fact's denial.

Consider the implications of this claim for the problem of imposing burdens on one person to benefit someone else. The attempt to balance the gains of one against the losses of another is ruled out by some people, and it is minimally viewed as needing a special justification by others. In general, the claim that the gains outweigh the losses, which would satisfy a utilitarian, will not satisfy those who think the boundaries between persons should be more strictly protected and harder to cross. One way to formulate respect for the boundary would be what Parfit (1984:337) calls the *Claim about Compensation:* Someone's burden cannot be *compensated* by benefits to someone else.

Parfit says we cannot deny the Claim about Compensation. Nevertheless, becoming Reductionists by abandoning the Non-Reductionist View can have two effects on the Claim about Compensation. In becoming Reductionists, Parfit argues, we come to believe that personal identity is not what matters in metaphysics or morals. Rather, psychological connectedness and continuity are what matter. Consequently, if certain parts of our lives are not well connected, we should treat them, in some respects, like different lives. This belief suggests we might extend the *scope* of the Claim about Compensation. It should apply within lives and not just between them. For example, we might think burdens imposed on the child are not compensated by gains to the adult.

The second effect of becoming a Reductionist may be that we change the *weight* we give the Claim about Compensation. If the unity of persons is less, so is their separateness. But if separateness involves less, it also has less moral significance. So the Claim about Compensation may be given less moral weight, perhaps none at all. Doing this may support the utilitarian view that we need no special justification to aggregate benefits and burdens across persons.

These conclusions clearly threaten a central assumption of the Prudential Lifespan Account. I have assumed that the boundary between persons *is* morally significant and that it is always problematic how we are to aggregate or balance benefits and burdens across those boundaries. Consequently, I sought the theoretical advantage that would result from reducing the interpersonal problem of justice between age groups to an intrapersonal problem of prudent allocation over the lifespan. In what follows, I cannot respond to Parfit's extended argument in favor of Reductionism and against Non-Reductionism in any detail. Instead, I shall suggest that there is an alternative way to view the results of some of his arguments against a Non-Reductionist View. On this alternative view, persons and personal identity may indeed "matter," though not for the reasons that underlie a Non-Reductionist View. My argument here draws on my earlier work (see Daniels 1979; for a similar critique of Parfit see White 1986 [unpublished]).

I must first review at least the outline of Parfit's attack on the Non-Reductionist View. On a Non-Reductionist View, a person is a separately existing entity, distinct from his brain and body. The classical example is the Cartesian ego. On a Reductionist view, a person's existence consists in the existence of a brain and body, and the occurrence of interrelated physical and mental events. On this view a person is distinct from his brain and body but is not a separately existing entity. Parfit (1984:211) compares persons to nations, as Hume did. A nation's existence involves the existence of its citizens, living together in certain ways, on its territory, and a nation is distinct from its citizens and territory; but a nation is not a separately existing entity.

These different views of the metaphysics of persons not surprisingly give rise to different views of personal identity. The Non-Reductionist views personal identity as an all-or-nothing, determinate matter of fact. It is a separate fact beyond the facts about physical and psychological continuity. The fact of personal identity holds in every case completely or not at all. The Reductionist believes personal identity involves only physical and psychological continuity, facts which can be described in an impersonal way. To the Reductionist, personal identity may not be determinable in certain cases: There may be no fact of the matter. Moreover, personal identity is not what matters—or should not be what matters—when we think about our relationship to future person stages. Only the facts about connectedness and continuity matter to our survival and how we should view its importance to us.

Parfit brings two main lines of evidence to bear against the Non-Reductionist View. One line draws on actual cases of people with split brains (severed corpora callosa) who experience two unified streams of consciousness. Since "ownership" by a person cannot explain the unity of these streams of consciousness, Parfit (1984:245ff) argues, we have evidence for the Reductionist View.

The other line of evidence rests on thought experiments and explores our beliefs about personal identity in cases we do not, and perhaps cannot, actually encounter. The Non-Reductionist believes that personal identity must always be a fact of the matter when we consider a relationship between two person stages. Parfit challenges this belief by imagining thought experiments in which there is no fact of the matter about whether the same person exists through certain transformations. The question, Is this the same person?, becomes empty in certain cases, a consequence that cannot obtain if the the Non-Reductionist View is true. For example, Parfit (1984:236) imagines a neurosurgeon gradually replacing bits of his brain with bits of Greta Garbo's. Early in the operation, Parfit remains Parfit. At the end, he has become Garbo. In the middle, there is a spectrum of cases in which there is no answer to the question about personal identity. Consequently, we cannot accept the Non-Reductionist View since it implies there is always a fact of the matter.

A further implication of the Non-Reductionist View, which it shares with common sense, is that personal identity is what matters to us when we contemplate the future. When we think about a future pain, personal identity seems to be what matters. Whether it is *my* pain rather than the experience of someone just like me seems to make all the difference to my concern about it. Parfit wants to undermine this belief that personal identity is what matters.

Suppose I know that at some point in the near future there will be a person stage B which resembles what I am like in important ways. B has access to many of my memories and can act on many of the intentions I now form. B is psychologically connected and continuous with me. In Parfit's (1984:215) terms, Relation R holds between me and B, and we may suppose it holds only between me and B (and there are no competitors to B). Parfit argues that I now know all the relevant facts for deciding whether a severe pain that B will experience should make me anxious. But the facts that matter here have to do only with Relation R.

The situation is different for the Non-Reductionist, who thinks that not all the facts are in when I know about Relation R. A Non-Reductionist might still want to ask a further question: Is B's pain *mine*. Am I the self or subject who will experience the pain? The Non-Reductionist believes there is some deep, further fact that underlies our answer to these questions. For example, is there a Cartesian ego which B shares with me?

Parfit uses thought experiments to bring the contrast between the two Views into sharp relief and to permit us to "test" the contrasting accounts. For example, suppose I have use of Mr. Spock's teletransporter. I step into it. It dissolves me into particles which are dispersed, but it transports information about those particles sufficient to reconstitute an exact replica of me on Mars (or next door). Is the replica me? If it experiences pain, is it my pain? Or have I died and does someone else, who is exactly like me, now live? Should teletransportation be described as death or transportation?

We might think we could just ask the replica. But this will not do. Suppose the teletransporter malfunctions and deposits two replicas elsewhere, or leaves me intact and deposits a replica somewhere else. Both survivors in this case will think they are me. On Parfit's view, in this case, I may be survived by both replicas, but I shall be identical with neither, since identity is a one-to-one relationship. But if the machine does not malfunction, then I have no deeper reason or further fact to think the one replica or survivor is identical with me than I do to think either of the two survivors is. The only facts that count in determining survivorship are facts about connectedness and continuity (Relation R), and these are the only facts that remain to determine any meaningful notion of personal identity. The deep further fact the Non-Reductionist requires is just not available.

I believe we can reject the Non-Reductionist position without accepting Parfit's claim that our concepts of persons and personal identity are reducible to only those facts about connectedness and continuity with which he is concerned. These facts may constrain what count as acceptable concepts of persons—our concepts of persons and personal identity must be compatible with facts about connectedness and continuity. But other facts may contribute to the concepts of persons that we do and should employ. There is another version of Reductionism which is compatible with the claim that personal identity matters. If this is true, we can accept Parfit's critique of the Non-Reductionist without accepting his conclusions about the lack of depth or importance of the boundary between persons. There is space here only to sketch the alternative view.

First, it might be important to note that facts about the degree of connectedness or continuity among the stages of a person—the strength of Relation R—may themselves depend on facts about how persons think about themselves over time and on facts about how other persons think about us over time. What kinds of intentions I form, for example, will affect the relations of connectedness and continuity. But what intentions I form may depend on how connected and continuous I want the stages of my life to be. Similarly, I may make a special effort to preserve certain kinds of memories or to erase others. These actions will affect Relation R over time. This is not to suggest that the facts underlying Relation R are all under my control, but to some extent some of them surely are. Even if the facts underlying Relation R are all that matter, they are not fixed as metaphysical bedrock; they are plastic. I can make myself into a more or less connected person over time, depending on what importance I attach to doing so. The importance of the degree of connectedness is not itself determined by the facts of connectedness—at least I may be able to make the facts conform to some prior importance I attach to having them obtain.

Nor should we think of my decisions to affect connectedness in these ways as arbitrary. They may rest on important facts about me and my relations to others. Some projects, which I may consider of great importance, may be realizable only in the long run. To undertake them would require that I think of myself as committed over the long run. I may have to form certain intentions and engage those around me to treat me as having those intentions, if I am to succeed. Success will depend on my cooperation with myself and others over time. In this way, what is valuable to me may determine who I must become.

For my own purposes, I require the cooperation of others in shaping my life, including its degree of connectedness and continuity. But social purposes will also require that I treat myself and that others treat me as having a certain degree of connectedness and continuity over time. If others are to work cooperatively with me, and we are to share the benefits of cooperative activities, then we may attribute great importance to promoting continuity and connectedness within the lives of persons. We will want to hold people accountable—both legally and morally—to commitments and responsibilities over time. Social structures that promote our viewing ourselves as the same person over time may thus be necessary for us to live together in certain productive ways. If this is true, then these facts may require us to attach importance to personal identity which does not derive from the facts of connectedness and continuity themselves. We may even idealize (fictionalize?) how connected and continuous persons are in order to carry on certain activities which require those fictions. Nevertheless, those idealizations may define what persons are, or at least what our beliefs about them have to be, if we are to plan our lives effectively and cooperate with others successfully.

Metaphysics may be the bedrock, but it underdetermines what kinds of structures—including persons—we can and *ought to* build on it. The "ought to" here may be prudential, capturing what rational or reasonable persons should do. Or it may be moral. Parfit's claim that only Relation R matters and that personal identity does not would have to be rejected. The position I am sketching, which implies what I have elsewhere (Daniels 1979) called the "plasticity" of persons, can be thought of as a Reductionist position. At least it does not depend on the belief that persons are separate entities, like Cartesian egos. But persons—their identity and survival—are not reducible merely to some given set of facts about connectedness and continuity, as Parfit has proposed. I am not interested here, however, in insisting that this account is properly described as Reductionist. I insist only that one can thus give a better account of how we do and should view persons than Parfit's. White (1986) has made a similar point using the terminology of "supervenience." The identity of a person is a fact that *supervenes* not just on a base of facts about connectedness and continuity but on a base that includes facts about the ways in which others view and treat us.

If my sketch of an alternative account is preferable to Parfit's, then the central assumption about the boundaries between persons that motivates my Prudential Lifespan Account is defensible. We need not believe in Cartesian egos, or defend them against Parfit, to think that the boundaries between persons are an important fact of deep moral significance. This account also has implications for my discussion of discounting for connectedness (see page 162ff.). Parfit argued that it was not irrational for me to think that my concern about my future well-being could be discounted according to my degree of connectedness with my future person stage, since connectedness is what really matters. But we have reasons to think personal identity important in ways that go beyond the facts about connectedness. Indeed, we might want to alter facts about connectedness in accordance with what we think important. People will then have good reasons not to discount for connectedness. Some of my earlier concerns in this chapter bear on these reasons.

In short, as I have attempted to demonstrate, both assumptions underlying the Prudential Lifespan Account can be defended against arguments of the sort Parfit offers. This does not prove that these classical assumptions are firm ground on which to build, but at least my construction does not rest on what we know to be sand.

References

Aaron, H. J. 1966. "The Social Insurance Paradox." *Canadian Journal of Economics and Political Science* 32 (August):371–377. Cited in Aaron 1982:76.

Aaron, H. J. 1982. *Economic Effects of Social Security*. Washington, DC: The Brookings Institution.

Aaron, H., and Schwartz, W. 1984. *The Painful Prescription*. Washington, DC: The Brookings Institution.

Achenbaum, W. A. 1983. *Shades of Gray: Old Age, American Values, and Federal Policies Since 1920*. Boston: Little, Brown.

Barry, B. 1965. *Political Argument*. London: Routledge and Kegan Paul.

Barry, B. 1978. "Circumstances of Justice and Future Generations." In B. Barry and R. I. Sikora (eds.), *Obligations to Future Generations*. Philadelphia: Temple University Press, pp. 204–248.

Binstock, R. H., and Shanas, E. (eds.). 1976. *Handbook of Aging and the Social Sciences*. New York: Van Nostrand Reinhold.

Bishop, C. 1981. "A Compulsory National Long-Term-Care Insurance Program." In J. J. Callahan and S. S. Wallack (eds.), *Reforming the Long-Term-Care System*. Lexington, MA: D.C. Heath, pp. 61–93.

Bishop, C., Karon, S. L., Greenberg, J. N., Cohen, M. A., and Wallack, S. S. 1986. "A Plan to Create Comprehensive Group Long-Term-Care Insurance for College and University Employees." Prepared for the Commission on College Retirement. Brandeis University.

Borgatta, E. F., and McCluskey, N. G. (eds.). 1980. *Aging and Society: Current Research and Policy Perspectives.* Beverly Hills, CA: Sage.

Boskin, M. J. (ed.). 1978. *The Crisis in Social Security: Problems and Prospects,* 2nd ed. San Francisco: Institute for Contemporary Studies.

Boskin, M. J., Kotlikoff, L. J., Puffert, D. J., and Shoven, J. B. 1986. "Social Security: A Financial Appraisal Across and Within Generations." [unpublished].

Boskin, M. J., and Shoven, J. B. 1984. "Concepts and Measures of Earnings Replacement During Retirement." Working paper No. 1360. National Bureau of Economic Research.

Boskin, M. J., and Shoven, J. B. 1986. "Poverty Among the Elderly: Where Are the Holes in the Safety Net?" National Bureau of Economic Research.

Brandt, R. 1979. *A Theory of the Good and the Right.* Oxford: Oxford University Press.

Buchanan, A. 1975. "Revisability and Rational Choice." *Canadian Journal of Philosophy* 3:395–408.

Buchanan, A. 1984. "The Right to a Decent Minimum of Health Care." *Philosophy and Public Affairs* 13(1):55–78.

Butler, R. N. 1975. *Why Survive?: Being Old in America.* New York: Harper and Row.

Callahan, D. 1985. "What Do Children Owe Elderly Parents?" *Hastings Center Report* 15(2)(April):32–33.

Callahan, J. J., and Wallack, S. S. (eds.). 1981. *Reforming the Long-Term-Care System.* Lexington, MA: D. C. Heath.

Caplan, A. 1981. "The 'Unnaturalness' of Aging—A Sickness Unto Death?" In A. L. Caplan, H. T. Engelhardt, Jr., and J. J. McCartney (eds.), *Concepts of Health and Disease.* Reading, MA: Addison-Wesley, pp. 725–737.

Children's Defense Fund. 1984a. *American Children in Poverty.* Washington, DC: Children's Defense Fund. Cited in Preston 1984:437.

Children's Defense Fund. 1984b. *A Children's Defense Budget.* Washington, DC: Children's Defense Fund. Cited in Preston 1984:440.

Cohen, G. G. 1978. *Karl Marx's Theory of History: A Defense.* Oxford: Oxford University Press.

Cromwell, J., and Kanak, J. 1982. "The Effects of Prospective Reimbursement on Hospital Adoption and Service Sharing." *Health Care Financial Review* 4(2):67.

Crystal, S. 1982. *America's Old Age Crisis.* New York: Basic.

Daniels, N. 1978. "Merit and Meritocracy." *Philosophy and Public Affairs* 7(3):206–223.

Daniels, N. 1979. "Moral Theory and the Plasticity of Persons." *Monist* 62(3):265–287.

Daniels, N. 1980. "Reflective Equilibrium and Archimedean Points." *Canadian Journal of Philosophy* 76(5):256–282.

Daniels, N. 1981. "Health Care Needs and Distributive Justice." *Philosophy and Public Affairs* 10(2):146–179.

Daniels, N. 1982. "Am I My Parents' Keeper?" *Midwest Studies in Philosophy* 7:517–540.

Daniels, N. 1985. "Family Responsibility Initiatives and Justice Between Age Groups." *Law, Medicine, and Health Care* 13(4):153–159.

Daniels, N. 1986. "Why Saying No to Patients in the United States is So Hard: Cost-Containment, Justice, and Provider Autonomy." *New England Journal of Medicine* 314:1381–1383.

Danziger, S., and Gottschalk, P. 1983. "The Measurement of Poverty Implications for Anti-poverty Policy. *American Behavioral Scientist* 26:746. Cited in Preston 1984:436.

Derthick, M. 1979. *Policymaking for Social Security.* Washington, DC: The Brookings Institution.

Doty, P. 1986. "Family Care of the Elderly." *The Milbank Quarterly* 64(1):34–75.

Dunlop, B. 1980. "Expanded Home-Based Care for the Impaired Elderly: Solution or Pipe Dream?" *American Journal of Public Health* 70:514–519. Cited in Doty 1986:56.

Dworkin, R. 1986. "Philosophical Issues Concerning the Rights of Patients Suffering Serious Permanent Dementia." Office of Technology Assessment. U.S. Congress [unpublished].

Elster, J. 1984. *Ulysses and the Sirens,* 2nd ed. Cambridge: Cambridge University Press.

English, J. 1979. "What Do Grown Children Owe Their Parents?" In O. O'Neill and W. Ruddick (eds.), *Having Children: Philosophical and Legal Reflections on Parenthood.* New York: Oxford University Press, pp. 351–356.

Estes, C. L. 1979. *The Aging Enterprise.* San Francisco: Jossey-Bass.

Estes, C. L., Newcomer, R. J., et al. (eds.). 1983. *Fiscal Austerity and Aging.* Beverly Hills, CA: Sage.

Featherman, D. L. 1983. "The Life-span Perspective in Social Science Research." In P. B. Blates and O. G. Brim, Jr. (eds.), *Life-span Development and Behavior,* vol. 5. New York: Academic Press, pp. 1–59.

Frankfather, D. L., Smith, M. J., and Caro, F. G. 1981. *Family Care of the Elderly.* Lexington, MA: Lexington Books.

Gauthier, D. 1986. *Morals By Agreement.* Oxford: Oxford University Press.

Gibbard, A. 1982. "The Prospective Pareto Principle and its Application to Questions of Equity of Access to Health Care: A Philosophical Examination." *Milbank Memorial Fund Quarterly/Health and Society* 60(3):399–428.

Gibson, R. M., and Fisher, C. R. 1979. "Age Differences in Health Care Spending, Fiscal Year 1977." *Social Security Bulletin* 42(1):3–16.

Goody, J. 1958. "The Fission of Domestic Groups Among the LoDagaba." In J. Goody (ed.), *The Developmental Cycle in Domestic Groups*. Cambridge: Cambridge University Press, pp. 53–91.

Goody, J. 1976. "Generations and Intergenerational Relations: Perspectives on Age Groups and Social Change." In R. H. Binstock and E. Shanas (eds.), *Handbook of Aging and the Social Sciences*. New York: Van Nostrand Reinhold, pp. 130–159.

Gordon, M. (ed.). 1978. *The American Family in Social-Historical Perspective*, 3rd ed. New York: St. Martins.

Harris, L. 1981. *Aging in the Eighties: America in Transition*. Washington, DC: National Council on the Aging.

Harsanyi, J. C. 1976. *Essays on Ethics, Social Behavior, and Scientific Explanation*. Dordrecht: Reidel.

Horowitz, A., and Dobrof, R. 1982. "The Role of Families in Providing Long-Term Care to the Frail and Chronically Ill Elderly Living in the Community. Final Report. Health Care Financing Administration Grant No. 18-P-97541/20-02. [unpublished]. Cited in Doty 1986.

Jonsen, A. 1976. "Principles for an Ethics of Health Services." In B. Neugarten and R. J. Havighurst (eds.), *Social Policy, Social Ethics, and the Aging Society*. Washington, DC: National Science Foundation, pp. 97–105.

Kagan, S. 1986. "The Present-Aim Theory of Rationality." *Ethics* 96(4):746–759.

Kane, R. L., and Kane, R. A. *A Will and a Way: What the United States Can Learn from Canada about Caring for the Elderly*. New York: Columbia University Press.

Kapp, M. B. 1978. "Residents of State Mental Institutions and Their Money (Or the State Giveth and the State Taketh Away)." *Journal of Psychiatry and the Law* 6(3):287–305.

Kingson, E. R., Hirshorn, B. A., and Harootyan, L. K. 1986. *The Common Stake: The Interdependence of Generations: A Policy Framework for an Aging Society*. Washington, DC: Gerontological Society of America.

Knox, R. 1984. "Fund Cuts are Linked to Infant Death Rise." *The Boston Globe* 225(May 24):1, 20.

Kreps, J. 1971. *Lifetime Allocation of Work and Income: Essays in the Economics of Aging*. Durham, NC: Duke University Press.

Kutza, E. A. 1981. *The Benefits of Old Age: Social-Welfare Policy for the Elderly*. Chicago: University of Chicago Press.

Laslett, P. (ed.). 1972. *Household and Family in Past Time*. Cambridge: Cambridge University Press.

Laslett, P. 1976. "Societal Development and Aging." In R. H. Binstock and E. Shanas (eds.), *Handbook of Aging and the Social Sciences*. New York: Van Nostrand Reinhold, pp. 87–116.

Leimer, D. R., and Petri, P. A. 1981. "Cohort Specific Effects of Social Security Policy." *National Tax Journal* 34(March):9–28. Cited in Aaron 1982:74.

Lesnoff-Caravaglia, G. (ed.). 1985. *Values, Ethics, and Aging*. New York: Human Sciences Press.

Longman, P. 1985. "Justice Between Generations." *The Atlantic Monthly* 255(6):73–81.

Manton, K. G., and Lui, K. 1984a. "The Future Growth of the Long-Term Care Population: Projections Based on the 1977 National Nursing Home Survey and the 1982 Long-Term Care Survey. Paper presented at the Third National Leadership Conference on Long-Term Care Issues. Washington, DC, March 7–9. Cited in Soldo and Manton 1985.

Manton, K. G., and Soldo, B. J. 1985. "Dynamics of Health Changes in the Oldest Old: New Perspectives and Evidence." *Milbank Memorial Fund Quarterly/Health and Society* 63(2):206–285.

Marmor, T. R. 1973. *The Politics of Medicare*. New York: Aldine.

McNeely, R. L., and Colen, J. L. (eds.). 1983. *Aging in Minority Groups*. Beverly Hills, CA: Sage.

Meltzer, J., Farrow, F., and Richman, H. (eds.). 1981. *Policy Options in Long-Term Care*. Chicago: University of Chicago Press.

Moffitt, R. 1982. "Trends in Social Security Wealth by Cohort." Paper prepared for the National Bureau of Economic Research Conference on Income and Wealth. Madison, WI, May 14–15. Cited in Aaron 1982:6.

Monk, A. 1979. *Age of Aging: A Reader in Social Gerontology*. New York: Prometheus.

Morgan, J. N. 1976. "The Ethical Basis of the Economic Claims of the Elderly." In B. Neugarten and R. J. Havighurst (eds.), *Social Policy, Social Ethics, and the Aging Society*. Washington, DC: National Science Foundation, pp. 67–69.

Morris, R., and Youket, P. 1981. "Long-Term Care Issues: Identifying the Problems and Potential Solutions." In J. J. Callahan and S. S. Wallack (eds.), *Reforming the Long-Term Care System*. Lexington, MA: D. C. Heath, pp. 11–28.

Myerhoff, B. G., and Simic, A. (eds.). 1977. *Life's Career—Aging: Cultural Variations on Growing Old*. Beverly Hills, CA: Sage.

Neugarten, B. 1974. "Age Groups in American Society and the Rise of the Young-Old." *Annals of the American Academy of Political and Social Science* 415:189–198.

Neugarten, B. (ed). 1982. *Age or Need? Public Policies for Older People.* Beverly Hills, CA: Sage.

Neugarten, B., and Havighurst, R. J. (eds.). 1976. *Social Policy, Social Ethics, and the Aging Society.* Washington, DC: National Science Foundation.

Nozick, R. 1974. *Anarchy State and Utopia.* New York: Basic.

Olson, L. K. 1982. *The Political Economy of Aging: The State, Private Power, and Social Welfare.* New York: Columbia University Press.

Olson, L., Caton, C., and Duffy, M. 1981. *The Elderly and the Future Economy.* Lexington, MA: D. C. Heath.

OTA. 1983. *Medical Technology and Costs of the Medicare Program.* Washington, DC: U.S. Congress, Office of Technology Assessment, OTA-H-227.

OTA. 1985. *Technology and Aging in America.* Washington, DC: U.S. Congress, Office of Technology Assessment, OTA-BA-264.

Palmore, E. 1976. "Total Chance of Institutionalization Among the Aged." *The Gerontologist* 16(6):504–507.

Parfit, D. 1973. "Later Selves and Moral Principles." In A. Montefiori (ed.), *Philosophy and Personal Relations.* London: Routledge and Kegan Paul, pp. 137–169.

Parfit, P. 1984. *Reasons and Persons.* Oxford: Oxford University Press.

Parsons, D. O., and Munro, D. R. 1978. "Intergenerational Transfers in Social Security." In M. J. Boskin (ed.), *The Crisis in Social Security: Problems and Prospects,* 2nd ed. San Francisco: Institute for Contemporary Studies, pp. 65–86.

Pegels, C. C. 1981. *Health Care and the Elderly.* Rockville, MD: Aspen.

Preston, S. H. 1984. "Children and the Elderly: Divergent Paths for America's Dependents." *Demography* 21(4):435–457.

Rawls, J. 1971. *A Theory of Justice.* Cambridge, MA: Harvard University Press.

Rawls, J. 1980. "Kantian Constructivism in Moral Theory." *Journal of Philosophy* 77(9):515–572.

Rawls, J. 1982. "Social Unity and the Primary Goods." In A. K. Sen and B. Williams (eds.), *Utilitarianism and Beyond.* Cambridge: Cambridge University Press, pp. 159–185.

Rawls, J. 1985. "Justice as Fairness: Political not Metaphysical." *Philosophy and Public Affairs* 14(3):223–251.

Rice, D. P., and Feldman, J. J. 1983. "Living Longer in the United States: Demographic Changes and Health Needs of the Elderly." *Milbank Memorial Fund Quarterly/Health and Society* 61(3):363–396. Cited in Soldo and Manton 1985:286.

Robertson, A. H. 1981. *The Coming Revolution in Social Security.* Reston, VA: Reston Publishing Co.

Russell, L. 1982. *The Baby Boom Generation and the Economy*. Washington, DC: The Brookings Institution.

Samuelson, P. 1958. "An Exact Consumption-Loan Model of Interest With or Without the Social Contrivance of Money." *Journal of Political Economy* 66(6):467–482.

Scanlon, T. M. 1975. "Preference and Urgency." *Journal of Philosophy* 77(19):655–669.

Schelling, T. 1986. *Choice and Consequences*. Cambridge, MA: Harvard University Press.

Shanas, E. 1979. "The Family as a Social Support System in Old Age." *The Gerontologist* 19(2):169–174.

Shanas, E., Townsend, P., Wedderburn, D., Friis, H., Milhoj, P., and Stehouwer, J. 1968. *Old People in Three Industrial Societies*. London: Routledge and Kegan Paul.

Sheppard, H. L., and Rix, S. E. 1977. *The Graying of Working America*. New York: Free Press.

Shoeman, F. 1980. "Rights of Children, Rights of Parents, and the Moral Basis of the Family." *Ethics* 911:6–19.

Sidgwick, H. 1907. *The Methods of Ethics*. London: Macmillan.

Soldo, B. J., and Manton, K. G. 1985. "Changes in the Health Status and Service Needs of the Oldest Old: Current Patterns and Future Trends." *Milbank Memorial Fund Quarterly/Health and Society* 63(2):286–323.

Sommers, C. H. 1986. "Filial Morality." *Journal of Philosophy* 83(8):439–456.

Spengler, J., and Kreps, J. 1963. "Equity and Social Credit for the Retired." In J. Kreps (ed.), *Employment, Income, and Retirement Problems of the Aged*. Durham, NC: Duke University Press, pp. 198–229.

State House Notes. 1984. 42:7.

Stenning, D. J. 1958. "Household Viability Among the Pastoral Fulani." In J. Goody (ed.), *The Developmental Cycle in Domestic Groups*. Cambridge University Press, pp. 92–119. Cited in Goody 1976.

Treas, J. 1977. "Family Support Systems for the Aged: Some Social and Demographic Considerations." *The Gerontologist* 17(6):486–491.

U.S. Bureau of the Census. 1983a. *Statistical Abstract of the United States*. Washington, DC: U.S. Government Printing Office. Cited in Preston 1984.

Vaupal, J. W. 1976. "Early Death: An American Tragedy." *Law and Contemporary Problems* 40(4):73–121.

Veatch, R. M. (ed.). 1979. *Life Span: Values and Life-Extending Technologies*. New York: Harper and Row.

Viscusi, W. K. 1979. *Welfare of the Elderly: An Economic Analysis and Policy Prescription*. New York: John Wiley.

Vogel, R. J., and Palmer, H. C. (eds.). 1982. *Long-Term Care: Perspectives from Research and Demonstrations*. Washington, DC: Health Care Financing Administration, U.S. Department of Health and Human Services.

White, S. 1986. "Metapsychological Relativism and the Self." [unpublished].

Yelin, E. H., Hencke, C. J., Kramer, J. S., Nevitt, M. C., Shearn, M., and Epstein, W. V. 1985. A Comparison of the Treatment of Rheumatoid Arthritis in Health Maintenance Organizations and Fee-for-Service Practices. *New England Journal of Medicine* 312:962–967.

Zook, C. J., and Moore, F. D. 1980. "High-cost Users of Medical Care." *New England Journal of Medicine* 302(18):996–1002.

Index

185